The Green Car Guide

The Green Car Guide

Paul Nieuwenhuis
Peter Cope
Janet Armstrong

First published in 1992 by
Green Print
an imprint of the Merlin Press
10 Malden Road, London NW5 3HR

ISBN 1 85425 076 0

Phototypeset by Computerset Ltd., Harmondsworth, Middlesex

Printed in England by Biddles Ltd., Guildford, Surrey on recycled paper

Contents

Acknowledgements

No work of this scope can come about without the help of a number of people. Even with three authors, each with our own areas of expertise, there are areas where others are more competent and we have used their services as much as possible. Others have helped by reading early drafts or developing ideas and concepts, while yet other people have supplied us with information for the book. In this respect, we would particularly like to thank the following people for their help: M.C. Armstrong, Bacton; Jeremy Boyce, Buxton; Philippe Cornelis, Harrold; Julie Houghton, Norwich; John Kendall, Sutton; John King, Norwich; Mercedes-Benz (UK) Ltd; Ruth Nieuwenhuis, Cardiff; Jan Ploeger, Delft; Kazmier Wysocki, Hackensack, NY; Heather Yaxley of Peugeot-Talbot.

Glossary

Carbon filter/trap A device, soon to be mandatory on all cars in the EC, which traps petrol fumes (VOCs) escaping from the fuel tank. These are then sent into the engine when the car is started up, thus eliminating waste and reducing VOC pollution.

Cd An aerodynamic unit of drag, i.e. the total resistance of an object moving through air. An upright square moving through air has a theoretical Cd of 1.0. Most cars are much better at between 0.3 and 0.4, with some even better than that.

Comité de Liaison de la Construction Automobile (CLCA) A European grouping of the national motor vehicle manufacturers' and importers' associations. ACEA or Association des Constructeurs Européens d'Automobiles is its successor. It was set up in 1990.

CV Commercial vehicle (van, truck/lorry, etc.). An **LCV** is a light commercial vehicle (van, light truck, pick-up truck, minibus, etc.). An **HGV** is a heavy goods vehicle.

CVT Continuously variable transmission is a type of transmission ('gearbox') without fixed ratios. Instead of having a limited number of four or five 'speeds', a CVT has an infinite number, thus enabling it to respond better to changes in engine output or road conditions.

Diesel cycle/Otto cycle The most popular types of internal combustion engine. The difference lies in the fact that in a diesel engine the fuel is injected into the combustion chamber at the top of the cylinder when there is a very high pressure in there, which causes the combustion (compression-ignition); in the Otto system the combustion is initiated by a spark from a spark plug (spark-ignition). A diesel engine has no spark plugs. These systems are named after their respective inventors, both Germans.

Drivetrain A commonly used term, which refers to the series of linked components that make the car move, i.e. engine, gearbox and other transmission components.

EC The European Community.

EGR Exhaust gas recirculation is a simple device, whereby a small amount of exhaust gas is diverted back to the engine and introduced into the inlet manifold. This leads to a reduction of the combustion temperature (due to the virtual absence of oxygen in the exhaust gas) and therefore a reduction in NOX emissions.

European Commission (CEC) The law-making body of the European Community. The Commission proposes EC legislation, but this has to be ratified by the Council of Ministers, which has the ultimate decision-making power. CEC stands for Commission of the European Communities.

Gallons When considering such things as US CAFE legislation, it is important to be aware of the difference between Imperial gallons – as used in the UK, Australia, New Zealand, etc. – and the US gallon. One Imperial gallon is the equivalent of 4.546 litres, while one US gallon is the equivalent of only 3.785 litres.

GRP Glass fibre-reinforced plastic is the most commonly used synthetic material for car bodies or truck cabs. It consists of a thin matting made of strands of glass, impregnated with certain types of plastic and left to harden. Through this bonding process, the glass fibre matting provides the otherwise weak plastic with great strength.

HPV Human powered vehicle. The most popular of these is of course the bicycle; however, in recent years the term has become increasingly used for a specialized type of vehicle, which has moved more towards being a kind of small car fitted with bicycle components. These HPVs are often based on two or three-wheeled reclining cycles.

LPG Liquified petroleum gas is a by-product of the petrochemical industry. In many cases it is still burned off; however, it is a suitable fuel for cars, which can be made to run on it with minor modifications and lower levels of harmful emissions than a petrol engine without catalyst. Although used in small quantities in most countries to power vehicles, it is particularly popular in Italy, the Netherlands and South Korea.

Octane The measurement of a fuel's antiknock properties, i.e. its ability to resist detonation during the engine's power stroke.

Ozone (O_3) Ozone is a molecule of oxygen, which consists of 3 oxygen atoms. It is less stable than molecular oxygen (O_2), but occurs under certain conditions. In the lower atmosphere or troposphere, it is considered a pollutant and is a product of air pollution gases. In the upper atmosphere or stratosphere, it helps to protect us from damaging radiation.

Parc The term parc, as car parc or truck parc, refers to the total number of cars or trucks in use in a particular place at a particular time.

'Provo'/Kabouter movement This was a radical hippy-based movement in Amsterdam in the late 1960s. The direct action Provo movement grew into the peace-loving Kabouter movement (kabouter means 'gnome' in Dutch), members of which were elected onto the local council. Many environmental ideas initiated at this time are still popular in the Netherlands.

PSA Peugeot S.A. is the holding company for several car producing companies. The Peugeot, Citroën and Talbot names are all part of this empire, which also comprises Peugeot-Talbot in the UK. PSA is one of the largest vehicle producers in Europe.

Regenerative braking A process used on electric vehicles, whereby the electric motors are used as generators when they are not driving the car, but the car is moving, i.e. coasting or when going downhill. This also produces a braking effect and is based on the principle that an electric motor is essentially the same as a generator, only you feed electricity into one to produce movement and you feed movement into the other to produce electricity.

Rpm Revolutions per minute is a term used in engine design. It refers to the number of revolutions the crankshaft makes every minute. Most engines can only operate within a limited range of rpm, hence the need for a gearbox to allow a reasonable road speed to be reached despite these limitations.

SMMT The UK Society of Motor Manufacturers and Traders combines the producers and importers of cars, vans and trucks as well as some suppliers.

Test cycle Emissions standards in Western Europe, Japan and the US are derived from the test cycle. Each model a manufacturer wishes to sell has to undergo the test cycle in order to determine whether it meets the required standards. This entails measuring the car's emissions as it is driven at predetermined speeds over a set distance, and is supposed to reflect typical driving conditions. The parameters of the test are different in each major car producing region, and consequently it is impossible to carry out a precise comparison between the emissions standards which prevail in each of these regions (although an approximate comparison can be made).

Unleaded petrol Petrol to which lead has not been added. No petrol is fully lead-free, as even unleaded petrol contains minute quantities of the element.

VOCs Volatile organic compounds are hydrocarbons which in gaseous form are considered an air pollutant risk. Examples are petrol fumes or fumes from solvents used in paint plants. Attempts are now being made to capture these. In several countries, VOCs from petrol stations are now captured from the car's fuel tank and returned to the fuel reservoir under the forecourt, as the tank is filled. (See Carbon filter/trap).

1: *The motor car and the environment*

The rise of the motor car

The motor car is an important part of modern life. Within a century it has moved from being an eccentric and unreliable luxury to providing an indispensable means of private transport for all sorts of people across the world. Many of us could not do without a car and certainly no major industry could function competitively without a fleet of trucks and vans to deliver goods and services to customers.

But the motor vehicle, for all its success, is still a menace to our health and to the environment. It pollutes the air, causes unacceptable noise levels, hampers child development, kills and injures people, animals and plants and gobbles up fuel and land in its path.

There are probably more than 500 million motor vehicles in the world and an increase of 70 million by the end of the century is predicted for the EC alone. Most of these are driven on Japanese, West European and North American roads. But as the Third World develops and Eastern Europe starts to prosper, more cars and more trucks will be required to prop up their growing economies. If we do not start to think seriously about how to control the motor vehicle population (and its consequences), it will be the human population that suffers.

Persons per vehicle in selected countries

	1988	*1989*	*1990*
USA	1.7	1.2	1.2
West Germany	2.1	1.9	1.9
Great Britain	2.7	2.2	2.2
Japan	4.0	2.2	2.1
Portugal	7.2	5.4	4.7
USSR	22.5	12.5	11.0
Turkey	45.0	30.0	24.5
China	1440.0	265.0	245.0

(Source: SMMT)

As developing countries increase their vehicle populations (statistics show that the rate of motorisation is growing fast) and approach the US and Western Europe in terms of vehicle density, pollution levels will increase further.

D-Day at last?

The 'Green revolution' has certainly made the car a prime target, and this is hardly surprising. The reluctance of manufacturers to mass-produce a 'clean' product has made the ordinary motor car a sitting duck. The strength of the transport lobby (a parliamentary lobbying committee made up of various organisations related indirectly or directly to the motor industry such as the AA, the SMMT and the RAC) has ensured up till now that roadbuilding and the growth of road transport has taken precedence, at least in the UK, over the promotion of environmental needs.

It looks, though, as if times are changing. Environmental groups and public interest in green issues have grown tremendously over the last few years. Groups such as Friends of the Earth and Greenpeace have rapidly risen from being fringe organizations with a reputation for crankiness to being an influential political lobby. The success of green parties in elections right across Europe backs this up, as people place increasing importance upon the preservation of the environment. The tide may be turning for the motor car as we know it. It is clear that consolidated action is needed by governments, motor manufacturers and petroleum companies to clean up cars or to present the car-buying public with a viable alternative solution – a mode of private transport that gives the individual freedom of movement while also satisfying the imperative of environmental protection.

Pollution and pollutants

No one can ignore the problem of motor vehicle pollution. It only takes a stroll into a busy city centre to appreciate the health dangers of exhaust fumes, but cars cause far more damage than just coughing or sneezing. The recent publicity given to the issue of lead in petrol has served to highlight the real danger behind motor vehicle pollution, but the truth is that the problem goes far deeper than just removing the lead.

During its lifetime, an average car probably travels some 100,000 miles, consuming some 2500 gallons of fuel and around 50 gallons of oil. In return it emits a potpourri of unpleasant substances into the air. Motor vehicles account for a large proportion of air pollution in most developed countries. It has been estimated that cars alone account for some 10,000 billion cubic metres of exhaust fumes every year across the entire world. In the UK road transport produces around 5 million tonnes of carbon monoxide, 800,000 tonnes of nitrogen oxides and 600,000 tonnes of volatile hydrocarbons, all of which are harmful.

The European Commission in Brussels has recently been drawing up laws that prevent manufacturers from selling cars that produce unacceptable levels of pollution. The general idea is that, by January 1993, all new cars will have to become 'clean' by conforming to these new rules.

The Commission is primarily concerned with three gases that are emitted in a vehicle's exhaust pipe. They are oxides of nitrogen, carbon monoxide and hydrocarbons. These represent a very real danger to our health if they are inhaled and they also place our environment in danger when they react in the air (or on surfaces) to produce other harmful substances.

The new limits will apply equally to diesel and petrol cars and should help to reduce the levels of these toxic fumes in our environment. But we should remember that many of the environmental problems associated with the motor vehicle still remain unsolved (and largely undiscussed). Exhaust gases consist of more than three substances and the motor vehicle is itself the result of a complex and environmentally damaging production process. Moreover, the environmental consequences of emitting just these three substances into the atmosphere are enormous and are only the start of a chain of damaging events.

Lead

Lead – in the form of organic compounds such as lead tetraethyl – has been used for many years as a petrol additive to enhance engine performance and to prevent knocking (or 'pinking'). When fuel is burned, the lead is passed into the atmosphere via the exhaust system. Lead is a particularly dangerous poison, as it accumulates in the body over time. It cannot be got rid of once in the system and is said to affect the development of young children. The EC has taken steps to reduce the amount of airborne lead by ensuring the availability of unleaded petrol in the twelve member states. In addition, all new cars are now obliged by EC law to run on unleaded petrol. Since a 60% reduction in the lead content of petrol was made in 1986, airborne levels of lead have already fallen by 50%.

But evidence shows that changing over to unleaded is by no means a panacea. Further sharp declines in lead levels are not likely to occur until unleaded petrol is more widely used in the UK. In 1988 the level of lead in the air actually rose as the increase in traffic volume exceeded the uptake of unleaded petrol. As more and more people use cars as a means of transport, the problem gets worse. Several factors may be to blame. Public transport has become inadequate, but increased wealth and the relative cheapness of modern cars have also meant that more people have found cars indispensable.

Is unleaded petrol a red herring?

While the rapid conversion of British motorists to unleaded fuel has been welcomed by many as a triumph for the environmentalists and has helped to soothe the troubled consciences of many drivers, lead in petrol has very little to do with the gradual destruction of the environment caused in part by cars. It is unfortunate in a sense that the movement towards unleaded petrol has been so successful in attracting publicity as it has unwittingly served to detract from problems that are potentially more serious. Motor manufacturers have been allowed considerable freedom in the way in which they design cars and have tended to ignore the environmental consequences of both building and running motor vehicles. In fact since the dawn of the motor car most manufacturers have striven to make their cars as fast as possible, often sacrificing fuel economy in favour of a few extra miles per hour. Most people will recall that it was not until the oil crises of the 1970s that manufacturers started to think seriously about saving energy.

Hydrocarbons

Hydrocarbons (HCs) are a group of chemicals that include petrol, oil and gas and also various solvents used in industry. They are volatile and evaporate quickly into the air if spilled. If breathed in they can be very harmful. Typical hydrocarbons used in the household include toluene, which is used as a solvent for paints or adhesives.

Road traffic is responsible for producing around 28% of hydrocarbon emissions in the UK. Around two thirds of this is emitted by exhausts as unburnt fuel and another third is caused by evaporation from petrol pumps or from vehicle fuel systems. Hydrocarbons react in sunlight with nitrogen oxides to produce photochemical oxidants, including PANS (peroxacetyl nitrates) which are associated with the petrochemical smog that forms in hot areas such as California, and ozone. Both are irritating to humans and animals and injure plants. Ozone is particularly dangerous in hot areas. On the ground it can be harmful to lung tissue and to the body's immune system. In the US ozone has been known as a problem for many years. Even low levels in the lower atmosphere can be harmful to active people such as children and athletes. Ozone is formed when ultraviolet light breaks down nitrogen dioxide into nitrogen oxide and atomic oxygen. The released oxygen atoms then combine with the oxygen molecules (O_2) that occur naturally in the air to form ozone (O_3).

Ozone does not tend to accumulate if it can react rapidly with nitrogen oxide to reform nitrogen dioxide and oxygen. But if hydrocarbons are present in the air the nitrogen oxide will form nitrogen dioxide without using up the excess ozone. This means that sinks of ozone tend to form in the atmosphere which can present a danger to health. Hydrocarbon emissions have to be controlled to reduce the risk of ozone formation.

Hydrocarbons include a group of chemicals known as PAHs (polyaromatic hydrocarbons) which include benzo(a)pyrene and fluoranthene. Studies show that they increase the rate of lung cancer and there is no known acceptable exposure level.

Nitrogen oxides

Road traffic is the biggest single source of nitrogen oxides (NOx) in the UK – 45% of the total. Nitrogen oxides consist of nitric oxide (NO), nitrogen dioxide (NO_2) and nitrous oxide (N_2O). Nitrogen dioxide affects people who have cardiac and respiratory weaknesses, nitrous oxide helps to deplete ozone in the upper atmosphere (where it is needed to keep out harmful sunlight) and may also contribute to the greenhouse effect. Furthermore, oxides of nitrogen combine with water vapour to form one of the components of acid rain.

Carbon monoxide

Perhaps the best known of the three exhaust toxins, carbon monoxide (CO) is formed when fossil fuels burn only partially (not all the fuel inside a car engine will be burned completely, and some will even escape unburnt). In the UK, road transport accounts for around 85% of all carbon monoxide emitted into the atmosphere. When inhaled, carbon monoxide reduces the oxygen-carrying capacity of the blood and can cause headaches, stress and respiratory problems. Carbon monoxide can also kill in a matter of minutes if inhaled in large enough quantities. The World Health Organization has issued guidelines about atmospheric levels of carbon monoxide, which are regularly exceeded in London. Levels are also still on the increase in the UK. Carbon monoxide has no immediate effects on the environment but can lead to increased levels of ozone in the lower atmosphere (troposphere).

Other harmful substances

Sulphur oxides (SOx)

Mainly sulphur dioxide (diesel engines produce roughly six times the sulphur dioxide given off by petrol engines). Sulphur dioxide is a major contributor to acid rain but the road transport contribution to total sulphur dioxide emissions is small.

Soot particulates

Diesel engines produce small particles of soot which may be carcinogenic or carry carcinogenic materials and are responsible for soiling the environment.

Aldehydes

E.g. acetaldehyde, formaldehyde. These are contained in small amounts in exhausts and irritate eyes, nose and throat. Petrol engines produce twice as much as diesels. With an exhaust catalyst aldehyde emissions are prevented.

Benzene
A carcinogenic substance emitted in large quantities by petrol engined cars (less so by diesels). Levels in European cities are already at a level that should cause some alarm.

Acid rain

One of the best known consequences of the motor vehicle is acid rain. Acid rain is a growing problem common in many European countries and has been linked both to an increasing vehicle population and to rapid industrialization. Apart from being atmospheric pollutants in their own right, nitrogen oxides (alongside carbon dioxide and sulphur dioxide) form a strong acid in the moist atmosphere which has the result of lowering the pH of rain, making it more acidic and more environmentally damaging. We should stress, however, that motor vehicles are not the only culprit here, nor the principal one.

The pH scale

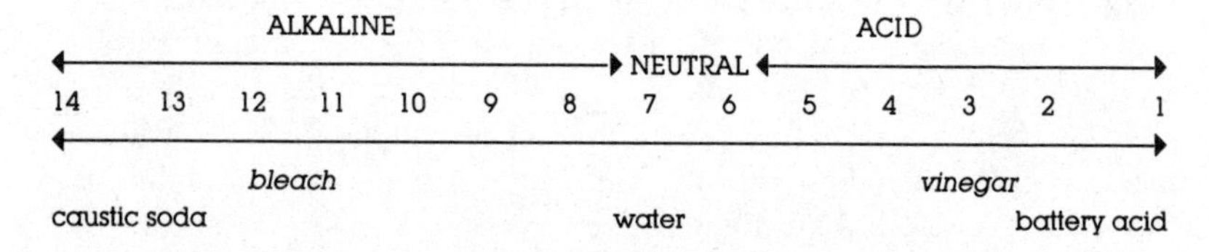

The pH scale (see diagram) is used to indicate the relative acidity or alkalinity of a substance. The lowest pH is 1 (which is very acidic) while the highest is 14 (very alkaline). Rain is naturally slightly acidic with a pH of around 5 to 5.6 but acid rain has been measured with a much lower pH in some areas of Europe. The important thing to remember with this scale is that it is logarithmic, so that one unit on the scale indicates a ten-fold increase or drop in acidity or alkalinity. So if the pH of rain drops from 5 to 4, it has become ten times more acid.

A more accurate term for acid rain would be acid deposition, since we should take account of both wet and dry deposition, both of which can damage the environment and both of which are caused by motor vehicles. Dry deposition is the fallout of nitrogen oxides and sulphur dioxide from the atmosphere. This type of deposition is usually deposited close to the source and typically includes soot on buildings, soil and crops. Wet deposition includes acid rain, snow, hail and mist and can be carried long distances in the atmosphere.

Acid deposition causes damage to freshwater life, to vegetation and to buildings and materials (it can even damage the paintwork on your car, causing discolouration and condensation to form on the surface). Decreases in the pH of lakes in the south of Norway have been measured from around

5.8 in 1940 to 4.6 in 1976/78. A pH of 4.5 or below means that fish become vulnerable and their populations decline. Declines in fish populations have obvious ecological consequences for human and animal populations that feed upon them.

Forest decline caused by acid deposition is also a problem, first detected in 1980. It affects many European countries, from Italy to Sweden (including Britain). The West German government has estimated that 50% of that country's forests have damaged trees. The damage entails disruption in growth, discolouration and droppage of leaves, and eventually the death of the tree itself. Acid rain has economic consequences too: loss of forest yield in the EC has been estimated at $300 million per year.

Acid rain also causes damage to buildings. Sulphur-based acids penetrate stonework, particularly limestone, where they convert the calcium carbonate to calcium sulphate, which crystallises and causes the stone to crumble. Corrosion also affects metal, stained glass, marble and paint. This type of damage can be very costly and may cost EC states in excess of $2 billion per year in repairs and maintenance.

Carbon dioxide and the greenhouse effect

Any process that involves the burning of fossil fuels produces carbon dioxide. Motor vehicles are no exception and account for 16% of the UK's human-made carbon dioxide emissions.

Carbon dioxide is a naturally occurring gas in the atmosphere that is produced both by the breathing of animals (including humans) and by the decomposition or burning of organic matter. It is also released by erupting volcanoes! Carbon dioxide is a vital part of life on this planet and is only one link in a complex cycle of events in which the vital element carbon is constantly recycled and shared between the atmosphere, the oceans and the land. It is used by plants to make food and oxygen. This means that a balance should always be in operation. The oceans act as a buffer should any minor imbalance occur: carbon dioxide dissolves in the seawater from the atmosphere.

The burning of fossil fuels releases into the air annually about 5% as much carbon dioxide as the respiration of all organisms. This means that bit by bit the natural balance is being tipped. Human activities are shifting organic carbon from fossil fuels into the atmosphere as carbon dioxide, but as the rate of carbon usage has increased, the flux into the oceans has not kept up. Only half to two thirds of the excess is currently being absorbed and the net result is that atmospheric carbon dioxide is increasing.

Increasing concentrations of carbon dioxide in the atmosphere are central to fears about the greenhouse effect. Carbon dioxide is one of the major greenhouse gases which act very much like the glass of a greenhouse in that they permit rays of sunshine to pass through them but absorb heat radiated outwards from the earth – this means that heat is preserved within the lower atmosphere. Around 30 gases produced by human activity have been identified as contributors to the greenhouse effect. Motor vehicles produce indirectly or directly a number of these including carbon dioxide (in the exhaust), CFCs (chlorofluorocarbons) used in the air conditioning systems of luxury cars, nitrogen dioxide and ozone.

Until such time as the motor industry finds a replacement for fossil fuels, there is very little chance of reducing the contribution it makes to the greenhouse effect. Successive oil crises have done comparatively little to remind people of the fact that crude oil supplies will not last for ever, and consequently research into alternative fuels for motor vehicles is still largely in its infancy. Obviously the ideal fuel for a car would be pure sunlight, but such an alternative would only be truly viable in warmer climates. The Brazilian government did a lot in the past to establish ethanol as its main source of fuel for motor vehicles, but this measure was more a result of economic desperation than of concern for the environment. Ethanol does reduce nitrogen oxides, carbon monoxide and hydrocarbons from car exhausts, but it still gives off carbon dioxide when it is burnt, so really in this context its contribution is fairly minimal. However, as the Brazilian ethanol is directly derived from sugar cane, the cane as it grows may absorb an amount of carbon dioxide roughly equal to that emitted as it burns.

Hydrogen is a serious contender as an alternative fuel despite its reputation as a rather dangerous chemical. This was illustrated by the R101 and Hindenburg accidents of the 1930s, when these hydrogen-filled airships caught fire with disastrous consequences. Some manufacturers have been considering the use of liquid hydrogen in an experimental context. BMW has developed a hydrogen powered car and Mercedes-Benz has engineered a hydrogen powered bus. The great advantage of using hydrogen is that the only by-product of its oxidation is water, which can be used to produce more hydrogen fuel.

CFCs

Luxury cars in particular make common use of air conditioning systems, which require the use of CFCs. CFCs (or freons as they are also known) have been used as refrigerants for many years, but fell into disrepute when it was discovered that they break down in the stratosphere to release chlorine which destroys ozone molecules. In Montreal in 1978, legal regulations for the gradual reduction in the use of CFCs were initiated (the Montreal Protocol), and by 1995 many manufacturers of CFCs will discontinue production. There are still problems in this area, though, as some of the CFC substitutes that can be used have not yet been tested to an adequate extent to ensure their harmlessness and are not available in large enough quantities to supply the requirements of large-scale motor manufacturing. Although manufacturers of refrigeration systems are eliminating CFCs from their systems, some of them are using 'transition' substances such as hydrofluorocarbons (HCFCs), which still damage the ozone layer. In fact, the amount of airborne chlorine doubled between the early 1970s and the late 1980s, and is expected to continue to rise throughout the twenty-first century.

In spite of these setbacks, manufacturers can still ensure that CFC emissions from air conditioning systems are minimized by, for example, making sure that no leaks can occur during everyday operation or during servicing and by using only recycled refrigerants in new systems.

Making environmentally friendly vehicles

Around 30 million motor cars and 13 million commercial vehicles are produced every year all over the world making use of enormous quantities of natural resources such as iron, aluminium, water, solvents (used in plastics and paints) and coal (used to make energy to drive heavy machinery). The resulting environmental burden on the world's resources is massive.

It is not enough just to clean up the harmful gases produced by cars and lorries. Environmentally sound, in automotive terms at least, means the optimization of emissions, minimal consumption of natural resources, avoidance of harmful substances and the recycling of used materials.

Large factories are by nature environment-unfriendly. They use enormous amounts of heat and energy (usually in an inefficient way) and take up large amounts of space. Motor manufacturers also tend to attract large numbers of suppliers to their vicinity, such as seat, battery and tyre manufacturers. The motor industry is a complex business. Cars require an enormous amount of componentry which in turn uses additional resources. The manufacture of batteries, for example, has been criticized for its use of harmful chemicals such as cadmium. The manufacture of paints uses large amounts of industrial solvents, notorious for their health effects, in addition to other noxious additives such as lead chromate.

It is only recently that the major motor manufacturers (in North America and Europe at least) have started to take an interest in the effects that their plants have on the immediate environment. In Sweden, Volvo claims to have taken measures towards reducing sulphur dioxide emissions by heating its Torslanda plant using surplus heat from a nearby BP refinery, and Volkswagen has been quick to publicize its energy efficiency programmes already in progress in Germany.

It is clear that we as individuals have to do a lot of thinking about the ways in which we travel around from day to day. We, after all, are the ones who spend the money that keeps the motor manufacturers in business. At the end of the day we should be the ones who make the decisions. If we decide to go on relying on fossil fuels and wasting other valuable resources in this way then we must accept the consequences. But we also possess the power to change things. If we want solar powered cars, then we can press the motor manufacturers to make them. Some people have already shown their concern by voting for the green parties and by supporting organizations that further the environmental cause. If we do not make a move in the right direction now, it may be too late.

2: *Exhaust emissions*

The issues of unleaded petrol and catalytic converters have aroused much confusion and argument, both in the media and among motorists. As more and more manufacturers make catalytic converters available on their cars and as virtually all UK petrol stations were offering unleaded fuel by 1990, motorists have become concerned about how these new developments are going to affect them. Much of the current confusion arises from the fact that many people believe that using unleaded petrol is enough; that this will make their car 'clean' and that they will not have to make any more changes to their everyday motoring habits. This is not the case.

The issue of lead in petrol is largely a separate problem from that of air pollution caused by substances like oxides of nitrogen and sulphur dioxides and has accordingly been legislated for separately. Rather than being caused by the combustion process itself, lead is a fuel additive which is expelled into the air via the exhaust system of a car. It is a cheap way of increasing octane levels (the alternative is further refining) of petrol, thus increasing knock resistance and allowing engines with higher compression ratios to be used. A relatively simple adjustment enables a car – with hardened valve-seats (as lead acts as a lubricant for these) – to run on unleaded petrol and unleaded petrol has become widely available in Britain following the introduction of EC legislation in 1986. All new cars now have to run on unleaded petrol. In 1986 there had already been a reduction in the amount of lead in the atmosphere, although this figure actually rose in 1988, probably due to an increase in the volume of traffic.

Lead emitted by UK road vehicles 1982-1989

(thousands of tonnes)

1982 - 6.8
1983 - 6.9
1984 - 7.2
1985 - 6.5
1986 - 2.9 (First year that unleaded petrol becomes available)
1987 - 3.0
1988 - 3.1
1989 - 2.6

(Source: Digest of Environmental Protection and Water Statistics, HMSO)

While these figures are, on the whole, encouraging, changing to unleaded is not the only answer. British motorists still have a long way to go before air pollution can be improved.

Lead is by no means the only harmful substance emitted by motor vehicles. In fact, if you were to breathe in exhaust gases you would die far more quickly from carbon monoxide poisoning than from lead poisoning. People who jog through congested cities in the morning are probably damaging their health more by inhaling vast amounts of carbon monoxide rather than improving their fitness through exercise.

Because of the fuel they burn and regardless of its lead content, motor vehicles are one of the major causes of air pollution in the world today. Road transport alone creates about 85% of the UK's carbon monoxide emissions. A very large proportion of the country's nitrogen oxide and hydrocarbon emissions are also caused by motor vehicles.

On average a typical 1.6 litre car will deliver 400 lbs of unburned hydrocarbons, 250 lbs of nitrogen oxides and over 5000 lbs of carbon monoxide into the atmosphere during its lifespan, assuming a total lifetime mileage of 100,000 miles. Bearing in mind that there are over 20 million cars in the UK alone, not to mention trucks and vans, this creates a huge environmental burden. Increasing traffic congestion also means that pollution levels are being worsened. Every year in the UK about 2 million new cars are sold and the total number of vehicles in use increases every year.

Vehicles in use in Great Britain
(excluding agricultural vehicles and two-wheelers)

1982	20,127,336
1983	20,642,517
1984	21,184,383
1985	21,656,190
1986	22,182,383
1987	22,990,750
1988	24,018,758
1989	25,092,685
1990	25,769,453
2000	35,000,000???

How exhaust fumes are formed

To understand how pollutants are formed by cars it is necessary first to look at the combustion process; the burning of a fuel/air mixture inside the car's engine. Petrol and diesel are both hydrocarbons; that is, they are made up of chains of carbon atoms to which numerous hydrogen atoms are joined. When the mixture is ignited, the oxygen in the air reacts with the fuel hydrocarbons to produce vast amounts of heat and energy which effectively drives the car. If this combustion process were complete, all the hydrogen and carbon in the fuel would react with the oxygen, producing only carbon dioxide and water as end products. Unfortunately this can only occur if fuel

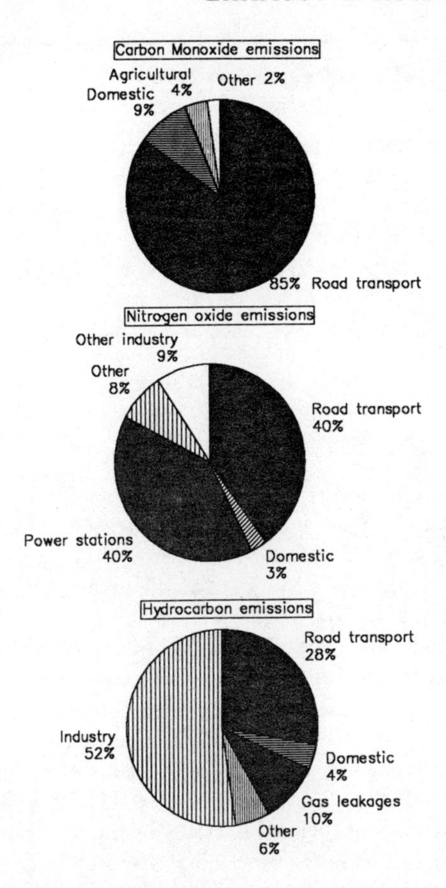

Emissions – percentage of total UK output by source

and air are supplied to the engine in exact proportions (14.7 times as much air as fuel). This is only possible in theory and the by-products of the combustion process that occurs in the average car engine are not only made up of these two substances. Not all the fuel burns in the engine and some burns only partially. This means that some hydrocarbons leave the engine unchanged while some carbon monoxide (another indicator of incomplete combustion) is also expelled.

The third major pollutant produced by motor vehicles is nitrogen oxides, which are formed when air (which contains not only oxygen but also nitrogen) burns at very high temperatures inside the combustion chamber of the engine.

Two schools of thought

Broadly speaking there are two schools of thought about the treatment of exhaust gases. Manufacturers can either decide to treat the harmful by-products of fuel combustion after they have left the engine or they can attempt to prevent their formation. The introduction of tough legislation by the EC has by and large led to the adoption of exhaust catalysts by motor manufacturers, which require less research and development, are quicker and cheaper to get to the marketplace and only treat pollutants after they have been formed.

Lean-burn technology

This is perhaps the most obvious method of controlling exhaust emissions as it is derived from fact that air and fuel, if burnt at the 'ideal' (or stoichiometric) ratio of 14.7:1, leave the lowest level of harmful by-products. The greater the air/fuel ratio, the greater the savings on fuel. Ratios of more than 20:1 have been achieved in some cases. Various motor manufacturers (such as Toyota, Peugeot, Ford and Rover) are carrying out research into this area but unfortunately the technology is largely still in its infancy and is unable to meet the highest emissions standards.

When a compressed fuel/air mixture is ignited inside the combustion chamber, a 'flame front' spreads out from the spark plug. As this flame front spreads out towards the walls of the cylinder, it is extinguished and leaves small pockets of unburnt fuel and partially burnt carbon monoxide which are passed out in the exhaust. These emissions can be reduced by burning a weaker mixture under higher compression. But the problem with this is that the higher temperatures required lead to a higher level of nitrogen oxide emissions. There is one exception to this; at a ratio of 21:1, the lack of fuel in the chamber causes a decrease in combustion temperature which reduces emissions of nitrogen oxides and also keeps carbon monoxide and hydrocarbon emissions at a low level. This is what is meant by 'true lean-burn'. Unfortunately the mixture required for this can only be achieved under certain ideal conditions (e.g. not under acceleration) and so even lean-burn engines of this type will not meet EC emissions standards.

Catalytic converters

By now most people will have heard of catalytic converters, or 'cats'. Perhaps we should start by dispelling one of the more popular misconceptions about catalytic converters. The new EC emissions laws do not *require* that all new cars must have a catalytic converter by the beginning of 1993. All that is stipulated by the new legislation is that all new small cars must comply with regulations that limit the amounts of carbon monoxide, hydrocarbons and nitrogen oxides that they produce under certain test conditions. How the manufacturers achieve this is up to them and they are not obliged to fit catalysts to their cars. Many manufacturers have adopted catalysts as a means of conforming to the new legislation, particularly as they provide the only truly feasible method of meeting the required standards.

The catalyst represents an easy and uncomplicated way of removing harmful by-products from the exhaust fumes of cars, but unlike lean-burn technology, the catalytic converter is by no means a new invention. In fact General Motors carried out the first significant research into their usage in the 1950s, but real advances did not occur until the late 1960s when the US first began to control emissions from cars.

The catalytic converter has become popular with motor manufacturers partly because it offers the most effective way of removing pollutants from the exhaust system. It is mounted close to the engine within the exhaust system so that all the exhaust gases have to pass through it before they reach the exhaust pipe. The catalyst itself is enclosed within a metal casing, but most catalysts have a matrix of ceramic passages called cells (of which there are around 400 to every square inch). The length of the passages of this matrix is around 3.5 metres, but the surface of the ceramic element is coated with a special rough surface 'washcoat' which is again coated with precious metals (such as platinum, palladium and rhodium) that increases the actual surface area of the catalyst to around the size of two football pitches!

Two-way (oxidation) catalysts

This is the simplest catalyst system available. It only treats two of the regulated pollutants – carbon monoxide and hydrocarbons – and uses the precious metals platinum and palladium to catalyse a reaction that converts these substances to water and carbon dioxide. The catalyst will only initiate the reaction if it is first heated to over 300°C, which is why it must be mounted close to the engine. The reaction itself also produces heat, so that the catalyst is actually working at temperatures between 400°C and 800°C.

Open-loop (or unregulated) three-way catalyst

Three-way open loop catalysts are common as 'aftermarket' add-ons to cars, meaning that they can be retro-fitted easily without having to make major adjustments to the engine or the fuel system of the car. A number of manufacturers such as Rover, Saab and Volkswagen currently offer three-way open loop catalysts on a variety of models as an aftermarket option.

The three-way catalyst uses platinum to initiate a reaction that oxidizes hydrocarbons and carbon monoxide to carbon dioxide and water (just like the oxidation catalyst). It also uses rhodium to initiate a reaction that reduces nitrogen oxides to simple atomic nitrogen. The term three-way simply means that three substances are treated by the catalyst.

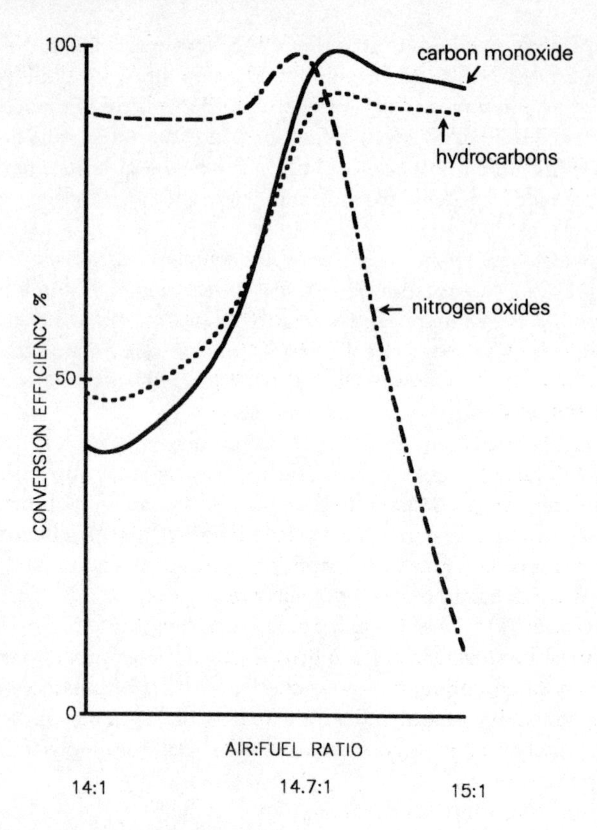

Air: fuel ratio and catalyst efficiency

Closed-loop (or regulated) three-way catalyst

The open-loop catalysts described above are the more simple methods of treating exhaust gases, but they are also less efficient and only remove some 70% of pollutants as a rule. The closed-loop system is more sophisticated and as a result is more effective. It works by making sure that the air/fuel ratio is maintained as close as possible to 14.7:1. This is because a catalytic converter will work best (at over 90% efficiency) when the mix is 'just right' so that there is adequate oxygen to carry out oxidation of carbon monoxide and hydrocarbons and not too much oxygen to inhibit the reduction of nitrogen oxides to simple nitrogen.

The closed-loop catalyst is able to maintain the air/fuel ratio at the 14.7:1 point – known as the stoichiometric point, or lambda – by using an oxygen sensor (a lambda sensor) that measures the amount of oxygen in exhaust gases leaving the engine. It does this by comparing the amount of oxygen in the exhaust gas with the amount in the atmosphere. If there is excess oxygen in the exhaust gas, the engine management system of the car will adjust the air/fuel ratio accordingly. This is the most efficient type of catalyst system, but it also requires additional componentry – such as an engine management system and fuel injection – which means that it has only been available as a rule on the more expensive models of car that already have most of the necessary technology.

The dual-bed system

This is rarely used now. It works by using one catalyst to deal with nitrogen oxides and another to deal with carbon monoxide and hydrocarbons. It has not caught on mainly because the engine has to run 'rich' (using more fuel in relation to air, which increases fuel consumption) so that the nitrogen oxide can be removed.

The drawbacks

Perhaps one reason why catalytic converters have been relatively unpopular so far is their tendency to cause power loss and a consequent decrease in performance. The industry average for a catalyst conforming to European norms is a 3% power loss while the corresponding loss for a cat-equipped car conforming to 1983 US norms is 8%. Manufacturers have tackled this problem by increasing engine capacity in order to compensate for the loss in power or by equipping cars with performance-enhancing modifications, such as turbochargers or multi-valve engines. For example, Volkswagen-Audi has increased the capacity of its 1.8 litre 16 valve engine by 200cc in its 'clean' catalyst-equipped form so that it still produces an equal amount of brake horse power (bhp) to its less 'clean' parent engine. The penalty is an increase in fuel consumption and thus carbon dioxide emissions. Power loss has also meant that manufacturers only tend to offer catalysts on models with larger engines.

Advocates of the lean-burn solution have warned that the cost of a catalyst to the consumer is considerable, especially if cars require a closed-loop system in order to meet the new European norms. Many manufacturers started offering catalysts as standard or as an option at around £300 to £400. This is common among the more upmarket models (such as Audi, BMW, Saab and Volvo) which tend already to have fuel injection and engine management systems. These items are less common on the cheaper, more popular models. There are still relatively few catalyst options open to the environmentally conscious car buyer who has a limited budget (although Volkswagen has been quick to advertise its catalyst-equipped small cars such as the Polo). It is at this end of the market that cost is likely to have greater implications for the consumer.

For a catalyst to work properly, exhaust gases must be able to reach the precious metal coating on the inner surface of the catalyst. If leaded petrol is used, the lead will rapidly clog the pores of the washcoat and will reduce the efficiency of the catalyst. This means that unleaded fuel must be used at all times in a catalyst-equipped car. The law requires that all vehicles equipped with catalytic converters must have a restrictor fitted to the fuel filler neck which will allow only the green-coloured unleaded pump to fit. Another implication of this is, of course, that most cars in the future will only run on unleaded fuel – meaning that perhaps leaded fuel will be phased out altogether, or will be very difficult to obtain.

The need for thorough and regular checks on the ignition system of a catalyst-equipped car is paramount. If the engine misfires this could lead to unburnt fuel entering the catalyst and burning, thus overheating the catalyst and completely destroying it. This will mean that car owners who are used to carrying out minor repairs and checks themselves will need to become more used to visiting their dealer who will be better prepared and better equipped to service the more complicated componentry. This will further increase motoring costs and will also require the motorist to become conscious at all times of the need to treat his/her car with care.

Diesel technology and emissions

Diesel cars will be subject to exactly the same emission norms as petrol cars when the new EC legislation comes into force at the beginning of 1993.

Diesel engines differ from petrol engines in that they are by nature lean-burn and burn fuel more efficiently, which means that they can produce on average less than 10% of the carbon monoxide and 25% of the hydrocarbons emitted by petrol engines equipped with a 90% efficient three-way catalyst. Nitrogen oxide emissions are also less than those given off by non-catalyst petrol engines. Because diesel engines are more fuel efficient they are cheaper to run and because they are built to withstand higher pressures they can endure longer mileages – which means that they last longer. Another advantage of diesel as a fuel is that it contains no lead additive, although sulphur additives are used to boost cetane levels (comparable to lead and octane levels in petrol) and this is the source of sulphur dioxide emissions from diesel engines. Only the oil companies can solve this problem.

In spite of the obvious environmental advantages of diesel engines, they do present a very real health risk in that they emit more smoke than petrol engines. In fact, the possibility that soot from diesel engines might have carcinogenic properties contributed to a collapse in the market for diesel cars in West Germany in the late 1980s. This in itself had useful consequences for the environment as manufacturers were led to develop cleaner versions of their diesel engines.

Diesel engines can be cleaned up in the same way as petrol engines by using catalysts to burn off excess pollutants. Volkswagen claimed to have developed what it calls 'the world's cleanest car' in the form of the 1.6 litre 60bhp 'Umwelt' Golf. The engine is derived from an existing 1.6 litre diesel engine to which a turbocharger and an oxidation catalyst have been added.

The turbocharger provides the engine with extra air which helps to reduce smoke production when the engine is under full load. The car is equipped with an oxidation catalyst that should help to eliminate over 50% of the unburnt hydrocarbons.

Mercedes-Benz has produced a 'cleaner' diesel engine which has been available in the UK since early 1989. The engine is said to comply with the new EC legislation. The reduction in emissions has been achieved by improvements inside the engine rather than by adding on extra components (such as catalysts). Fuel is injected more accurately into the combustion chamber so that combustion is more complete, leaving less unburned hydrocarbons and particles of soot. The company has also introduced a version of the same engine into the German market, accompanied with a turbocharger and an oxidation catalyst, so that emissions are still lower.

European car manufacturers have banded together to form IDEA (Integrated Diesel European Action), which also includes a number of academic institutions. The intention of the group is to research and examine all the processes involved with diesel engines and to produce an engine which emits fewer harmful substances and which optimizes efficiency.

Manufacturers are attempting to reduce particulate emissions by developing filter systems that effectively trap soot particles in the exhaust system and prevent them from leaving the exhaust and escaping into the air. The problem with this technique is that soot particles tend to accumulate in the filters and eventually block them completely. This leads to back-pressure on the engine from the exhaust gases which results in power loss and increased fuel consumption. Particles can be burnt off inside the filter if the temperature of the exhaust gases is high enough but this is not always possible if a car is only driven around town or at low speed. Some manufacturers are developing systems that encourage the soot particles to ignite at lower temperatures. This is often done by mixing an additive with the fuel.

3: *Government action on emissions*

California leads the way

Legislation to limit vehicle exhaust emissions was first enacted in California, which has the highest level of car ownership in the world. This was in the mid-1960s. Shortly afterwards, in 1970, the US introduced its own legislation at federal level, in the form of the Clean Air Act. This required the Environmental Protection Agency (EPA), America's environmental watchdog, to set air quality standards for sulphur dioxide, particulates, carbon monoxide, low level (tropospheric) ozone, nitrous oxides, and lead. The Clean Air Act was amended in 1977 and is still the basis of US efforts to reduce atmospheric pollution. Not all the standards set by the Clean Air Act have yet been met, and nine cities in particular – Los Angeles, San Diego, Houston, New York, Philadelphia, Hartford, Baltimore, Chicago and Milwaukee – failed to meet the end of 1987 deadline for ozone compliance.

Californian exhaust emissions standards are the strictest in the world and there is no doubt that emissions from individual cars have been reduced substantially since the introduction of controls. According to data from the US auto industry, hydrocarbons and carbon monoxide had been cut by 96% per car by 1989, while nitrous oxides had been reduced by 76%. However, this refers to the standards that each new car has to meet, and should not be confused with total emissions from all cars on the road. Total emissions are determined not just by the standards that cars have to meet, but by the number of cars that are in use. There has been a notable increase in car mileage since the 1960s, and it is this which has perpetuated the problem of smog in some major American cities, Los Angeles in particular.

The standards currently in force in the US are known as the US 1987 standards, which are essentially those standards introduced in 1983 with the addition of limits on diesel particle emissions. Since 1983 there has been no change in the limits set for the three major gaseous emissions, but legislation passed by Congress in November 1990 will lead to the introduction of stricter standards during the 1990s and beyond.

US car emissions standards for new cars (grams per mile and reduction since introduction)

	Hydrocarbons		Carbon monoxide		Nitrogen oxides	
Model year	*Grams*	*Reduction*	*Grams*	*Reduction*	*Grams*	*Reduction*
Pre-control	10.6	—	84.0	—	4.1	—
1968-71	4.1	62%	34.0	60%	NR	—
1972-74	3.0	72%	28.0	67%	3.1	24%
1975-76	1.5	86%	15.0	82%	3.1	24%
1977-79	1.5	86%	15.0	82%	2.0	51%
1980	0.41	96%	7.0	92%	2.0	51%
1981-82	0.41	96%	3.4	96%	1.0	86%
1983-92	0.41	96%	3.4	96%	1.0	76%

(Source: MVMA)

New US standards

Clean air legislation which had been making its way through Congress since the summer of 1989 was finally passed in November 1990. A bill introduced by the Bush Administration was amended in the House of Representatives Energy and Commerce Subcommittee in October 1989 and received the assent of the whole House in May 1990. This contained tougher exhaust emissions standards than President Bush had originally advocated. It called for a 40% reduction in emissions of hydrocarbons and a 60% reduction in emissions of nitrous oxides from each new car, to be phased in between model years 1994 and 1996. This would be followed by a further halving of emissions in 2004.

Competing legislation was passed by the Senate in April 1990, containing similar standards to those accepted by the House of Representatives but slightly earlier introduction dates (1993 and 2003). There followed a period of mediation between the chambers of Congress to determine the final bill which would be submitted for presidential assent. The resulting legislation contained the following provisions:

- A 60% reduction in nitrous oxide emissions (to 0.4 g/mile) and a 40% reduction in hydrocarbon emissions (to 0.25 g/mile) from each new car, commencing in the 1994 model year.
- A further halving of emissions in the 2004 model year, but only if the earlier reductions had failed to allow the country's six smoggiest cities to reach the required air quality standards.
- Certification by manufacturers that their anti-pollution equipment can last for ten years/100,000 miles, an eight-year/80,000 mile warranty for catalytic converters and on-board computers, and the fitting of cars with gauges which alert the driver if emissions control equipment is not working.
- The use of canisters which capture fumes during refuelling (providing that a US Depart of Transportation study finds them safe).

The legislation also provided a mandate for fuels which are cleaner than the currently available petrol.

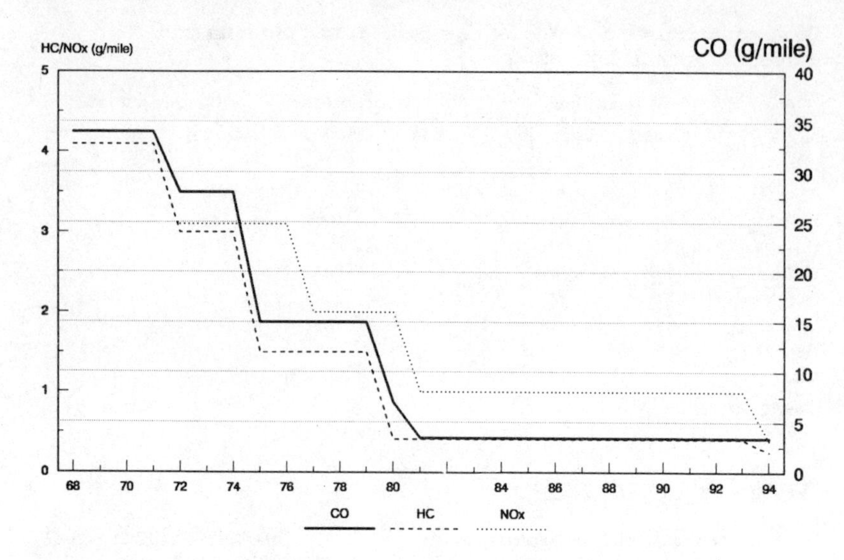

US emissions limits, 1968-94

The US exhaust emissions standards for the mid-1990s are identical to those already scheduled to enter force in California and eight northeastern states in 1993. However, true to the tradition of this state, California has recently rescheduled even stricter limits for progressive introduction over a period extending beyond the end of the century. The rationale behind this action is that the beneficial effects of the 1993 standards would eventually be overwhelmed by increases in the total usage of vehicles and thus prevent ambient air quality standards from being attained, especially in the Los Angeles area.

Under the latest Californian ruling, the ultimate standards for carbon monoxide and nitrous oxides will be the same as those embraced by Congress, but the hydrocarbon standard will eventually be as low as 0.07 g/mile, half the amount envisaged by the federal legislation.

The Californian regulation divides vehicles into four categories – transitional low emission vehicles (TLEV), low emission vehicles (LEV), ultra-low emission vehicles (ULEV), and zero emission vehicles (ZEV) – each of which will have to take a larger share of new car sales between the 1994 and 2003 model years. TLEVs will have to meet a hydrocarbon standard of 0.125 g/mile, while LEVs will have to reach 0.075 g/mile and ULEVs 0.04 g/mile. The aim is to achieve a sales-weighted fleet average for hydrocarbon emissions, which will be reduced progressively during the period in question as the proportion of less polluting cars sold increases.

More effective catalytic converters may allow some of the new standards to be met, but as they become stricter towards the end of the 1990s it is probable that vehicles will have to use cleaner fuel than is currently

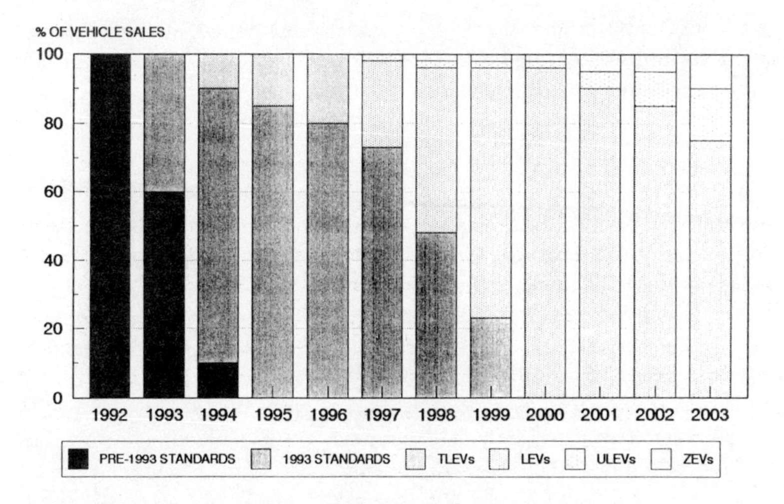

Californian exhaust emissions standards 1992-2003

available. This may be 'reformulated gasoline' or an alternative fuel such as methanol.

However, as their title suggests, the zero emission category will have essentially to be free of exhaust emissions, and means that they will have to be powered by electricity. Under the new Californian ruling, this category of vehicle should account for 2% of new car sales by 1998 and 10% by 2003. Furthermore, the South Coast Air Quality Management District, which includes Los Angeles, wants ZEVs to account for practically all sales by 2010.

Other countries adopt US-equivalent standards

Standards equivalent to those currently in force in the US have been adopted by a number of countries over the last few years. This is particularly the case in Europe, where Austria, Norway, Sweden and Switzerland now enforce US standards, with Finland following in 1992. The Danish parliament, not wanting to be left behind by its Scandinavian neighbours, voted in 1989 to introduce mandatory US standards from October 1, 1990. This made Denmark the first member state of the EC to opt for US standards.

Outside Europe, US standards have been introduced by Australia and South Korea, and both Brazil and Mexico plan to adopt them during the 1990s (in Mexico catalytic converters have effectively been required on new cars from the beginning of the 1991 model year). Japan enforces exhaust emissions limits which are thought to be broadly similar to those in the US, though measured in a different way.

Japan

The first Japanese exhaust emissions regulations were introduced in 1966, but have been tightened progressively since then. Regulation of hydrocarbon emissions from motor vehicles began in 1970, while control of nitrogen oxides, which the Japanese consider to be a particularly urgent problem, was commenced in 1973 and reinforced in 1975, 1976 and 1978. According to the Japan Automobile Manufacturers' Association (JAMA), the 1975 requirements represented a 95% reduction in carbon monoxide and a 96% reduction in hydrocarbons per vehicle, while the 1978 limit on nitrous oxides resulted in a 92% reduction from the pre-control level.

Japanese car exhaust emissions standards since 1975 (grams per kilometre)

Year	*Hydrocarbons*	*Carbon monoxide*	*Nitrogen oxides*
1975	0.25	2.10	1.20
1976	0.25	2.10	0.60 (0.85)*
1977	0.25	2.10	0.60 (0.85)*
1978-1990	0.25	2.10	0.25

*Figures in parentheses refer to cars with weight exceeding 1000 kg.
(*Source: Japan Automobile Manufacturers' Association*)

Diesel-powered cars, which at present account for only a very small proportion of the Japanese market, are allowed to emit a higher level of nitrogen oxides than their petrol or LPG-driven counterparts, but this is to change by the mid-1990s. On 1 September 1995 all diesel cars produced in Japan and sold in the domestic market will have to emit no more nitrogen oxides than petrol-powered cars (new models have to comply from October 1, 1994), and imported diesel cars will have to achieve the same standard by 1 April 1996.

In addition, a diesel particle standard of 0.34 g/km will be introduced on the same dates, along with a 20% reduction in emissions of black smoke. There may also have to be a significant cut in the sulphur content of diesel oil, so that catalytic converters can work effectively.

European developments

For the most part, Europe has lagged behind the US, Japan and some other countries in cutting permitted levels of toxic gases from car exhausts. This is not to say that there have been no controls on emissions; on the contrary, limits were introduced in 1971 and have been made stricter on several occasions since then. It is simply that, in the majority of West European countries, the standards have never been as rigorous as those enforced in the US and Japan.

The West European emissions limits were drawn up by the United Nations Economic Commission for Europe (UN/ECE), and accepted by the EC. In 1980 the fourth amendment to the standards was adopted by the UN/ECE, and the EC belatedly enshrined these new limits in 1983. By this time, however, a number of European countries were contemplating going their own way on the emissions question, and Sweden became the first to adopt its own national standards, based on the controls then in force in the US.

In the same year the EC decided to work towards further measures to combat air pollution, and in 1984 the European Commission proposed tighter emissions limits. The proposed controls were not stringent enough to satisfy some countries, notably Denmark and Greece, but in July 1985 the environment ministers of the then ten member states agreed on a formula which divided cars into three engine bands – less than 1.4 litres, 1.4 to two litres, and over two litres – with different emissions limits for each group.

In the large car (above two litres) group, the proposed standards were considered to be close to the US limits, and would necessitate the fitting of catalytic converters. For the medium category, lean-burn engines without catalysts were thought to be sufficient to meet the standards, and in the case of small cars existing engines might conform without significant modification.

The new standards were to be introduced over a period between 1988 and 1993, beginning with cars of above two litres engine capacity, but a lot of argument ensued between countries which thought the new limits to be sufficient and those which wanted US-equivalent controls. Furthermore, West Germany and the Netherlands went ahead and provided tax incentives for cars conforming to the proposed standards in advance of the anticipated dates of introduction, hoping thereby to accelerate sales of 'clean' cars.

Final agreement on all the proposed limits was not reached until November 1988, but by then pressure was beginning to build up for the introduction of US-equivalent standards throughout the EC. In West Germany, where events appeared to be moving particularly quickly, the environment minister, Klaus Töpfer, said early in 1989 that his country's motor industry should only make cars equipped with catalytic converters or diesel particle filters, and his statement was followed by an announcement by the conference of regional environment ministers that all new cars registered in West Germany should be fitted with a three-way catalytic converter from October 1991.

By this time, opinion in the European Commission was catching up with those countries which were supporting the introduction of US-equivalent standards. The Environment Commissioner, Carlo Ripa di Meana, proposed that US-equivalent standards be adopted in the EC from the beginning of 1993, and in April 1989 the European Parliament provided support for his position by voting for the introduction of US-equivalent standards for all new cars with engines of 1400cc and below by this date.

Given that EC member states had struggled for over three years (1985-88) to reach agreement on less stringent standards, it came as some surprise when, in June 1989, the EC environment ministers agreed that US-equivalent standards should be mandatory for new cars with engines of 1400cc and below from 1 January 1993. The mandatory nature of the standards represented a departure from earlier agreements, which had been optional in character – member states had had to allow manufacturers to sell cars which met the accepted standards, but they were not obliged to enforce them to the exclusion of cars which did not conform.

One element of the June 1989 agreement was that EC member states could provide tax incentives for the purchase of new cars fitted with catalytic converters, involving subsidies of up to 85% of the cost of the necessary equipment. This ratified existing practice in some countries, such as West Germany and the Netherlands.

Denmark voted against the new standards on the grounds that they were still not sufficiently stringent, and Greece, which had wanted US-equivalent standards to be introduced earlier in order to combat the severe pollution problem in Athens, voted against them because the tax incentives it was providing for catalyst-equipped cars were greater than those that the EC was prepared to accept.

On the other hand, France and Spain had been opposed to the new small car standards because they considered them too severe. In France, Peugeot was particularly hostile, considering the recourse to catalyst technology too hasty and damaging to the development of alternative solutions; in particular, it believed that the necessity to fit catalysts would compromise its work on lean-burn engines. The reasoning behind Peugeot's criticism of catalytic converter technology was shared in principle by a number of other European car manufacturers and by the industry's representative association, the CLCA. However, some of these producers, such as Fiat and Renault, had resigned themselves to the adoption of catalytic converters, and this helped to make the June 1989 agreement possible.

A few months after this agreement, the European Commission proposed that US-equivalent exhaust emissions standards be extended to cars of more than 1.4 litres, so that by the beginning of 1993 all new petrol-engined cars would effectively have to be fitted with three-way catalytic converters. The environment ministers accepted the Commission's proposals in July 1990, but the European Parliament then added a series of amendments, which resulted in a modified proposal from the Commission. This was adopted by the environment ministers in December 1990.

Apart from ensuring that all new cars will have to conform to US-equivalent exhaust emissions limits, the December 1990 agreement – as in the case of the small car agreement of June 1989 – allows governments to grant tax incentives up to a maximum of 85% of the cost of fitting a catalytic converter, provided that this is carried out before it becomes a legal requirement.

Furthermore, a new test cycle for the measurement of exhaust emissions is to be introduced. The existing test covers 4 km and includes only urban driving conditions at a maximum speed of 93 km/h; consequently it is considered to be a incomplete reflection of European driving conditions. The new test will retain the current procedures, but will add a new 7 km stretch involving out-of-town driving at speeds of up to 120 km/h (the Extra-Urban Driving Cycle).

Another measure that is likely to be introduced is a durability test for catalytic converters, similar in principle to that which is already enforced in the US. The test will be carried out after 80,000 km of use (this distance representing the average lifespan of a car), and will allow for some engine deterioration. The Commission had planned an initial test after 30,000 km at which point cars would still have to meet the stipulated emissions limits despite any deterioration of the engine, but this was dropped in the modified proposal.

By the beginning of 1993, then, emissions standards in the EC will be very close, if not quite identical, to those that were introduced in the US ten years earlier. There will be a further revision in the middle of the 1990s, shortly after the US standards themselves become more stringent, followed by another at the beginning of the next century.

The rise of unleaded fuel

The adoption of catalytic converter technology across the entire engine range is going to necessitate the use of unleaded fuel by all new petrol-engined cars within the next few years. Before this became apparent, however, the problem of lead in petrol was already a separate issue, lead being considered a health threat in its own right.

This was recognized at European Community level, which in March 1985 passed a directive (85/210/EEC) requiring member states to ensure the availability and balanced distribution of unleaded fuel having minimum octane ratings of 95 RON and 85 MON from October 1 1989. Furthermore, deadlines were set which mean that by now all new cars have to be capable of running on unleaded fuel – for cars of over two litres, the specified dates were October 1988 (new models) and October 1990 (new registrations), and for cars of two litres or less, the dates were October 1989 (new models) and October 1990 (new registrations).

All EC countries have introduced tax incentives which favour the purchase of unleaded fuel over the leaded variety. In the UK, an early champion of the cause of eliminating lead from petrol, the tax on unleaded fuel was reduced three times between 1988 and 1990, and it is now several pence cheaper per litre than leaded four-star petrol.

As a result of tax incentives and growing availability on petrol station forecourts, sales of unleaded fuel have risen sharply in a number of member states. Growth has been particularly swift in the UK and France – in the UK the market share of unleaded rose from a mere 5% at the beginning of 1989 to over 40% by the middle of 1991. In France, the market share of unleaded petrol was negligible in January 1989, but was approaching 25% by mid-1991.

Germany, however, has been the pioneer as far as sales of unleaded are concerned. As early as 1988, sales of unleaded fuel accounted for over 40% of the (West) German petrol market, and by the middle of 1991 the figure was approaching 80%. There were two main reasons for this state of affairs. Firstly, Germany was the first EC country to ban sales of leaded regular grade (two-star) petrol, which forced owners of Germany's large parc of low compression cars to opt for Eurograde unleaded. Secondly, from the late 1980s the vast majority of petrol-engined cars bought in Germany have, as a result of tax incentives and the actions of domestic car manufacturers, been equipped with catalytic converters, which require unleaded fuel if they are to function effectively.

Despite the fact that unleaded fuel can hardly be regarded as environmentally friendly, the actions of the EC and individual national governments mean that unleaded petrol is achieving a remarkable success and will continue to shrink the market share of leaded fuel until, possibly by the end of the 1990s, leaded is all but eliminated from the scene.

Diesels

The EC has also introduced new standards for diesel particle emissions, which came into force in October 1989 for new models and October 1990 for all new registrations. These standards are less strict than those enforced in the US, and are applied only to indirect injection diesels, as direct injection diesels, which tend to be less clean, were granted a reprieve until 1994. However, as in the case of gaseous emissions, the EC is catching up with current US practice, for in December 1990 the environment ministers accepted limits for diesel particles (0.14 g/km) which will, from the beginning of 1993, bring the Community's standards into line with those in the US.

There may be a further tightening up of particle emission limits in 1996, partly at the behest of the German government; a standard of 0.08 g/km has been mentioned, though even this is less severe than the standard which may ultimately be introduced in the US.

Despite the fact that diesel cars are in some respects less polluting than petrol-engined cars, they have been singled out in some countries as particularly damaging to air quality. This was the case until recently in (West) Germany, where a decision was made in the latter half of the 1980s to ban diesel cars from the road when smog reached a certain level. Furthermore, guidelines were published which stated that public authorities should purchase diesel cars only in substantiated exceptional cases, and diesels were excluded from the tax concessions granted to low polluting vehicles. However, by the beginning of the 1990s the German government was altering its position, and the tax incentives were restored. A statement was issued to the effect that diesel cars were not as hazardous as had been thought, and in consequence sales began to recover.

The Swedish government has been considering immediately phasing in a ban on diesel cars from Stockholm's streets, whereas petrol-engined vehicles would only be prohibited in about 2000-2005. Further afield, in Hong Kong, the government has been contemplating a ban on imports of diesel-engined vehicles, which constitute up to half of total vehicle mileage in the colony. The prohibition of high-sulphur fuel oil is also being planned there.

Alternative fuels: the key to ending urban air pollution?

At present there is no government-inspired policy in Europe, the US or Japan which favours the manufacture or purchase of alternative fuel vehicles on a large scale, although the US could be the first to introduce an alternative fuel programme.

The original clean air bill drafted by the Bush Administration in 1989 specified that manufacturers should sell 500,000 alternative fuel vehicles in 1995 and a million a year from 1997 to 2004. A watered down amendment emerged from the House of Representatives Energy and Commerce Subcommittee in autumn 1989, stating that manufacturers would only have to prove that they could make such vehicles in the necessary volumes, but after further compromises, the bill adopted in November 1990 stipulates that the motor industry will have to manufacture cars and light trucks which run on cleaner fuels, beginning in the 1996 model year. In addition, cleaner fuels will have to be sold in areas of deficient air quality (essentially the nine smoggiest cities in the US, referred to earlier in this chapter), and car fleets in the 24 smoggiest cities will have to use cleaner fuels from the 1998 model year.

Cleaner fuels include reformulated gasoline as well as alternative fuels such as methanol, natural gas or ethanol; they might also encapsulate flexible-fuel vehicles running on a blend of gasoline and methanol. It is uncertain as yet which forms of cleaner fuel will be adopted, but the US motor industry prefers reformulated gasoline, which would not require significant engine modifications. It requires fewer (harmful) additives to achieve a given octane rating, but costs a lot more to produce.

In Brazil, ethanol-powered cars have been in use since the 1970s, when the government encouraged alcohol as a fuel in order to reduce dependence on imported oil. Around half a million cars are powered by ethanol, which is obtained from sugar cane (70 litres of ethanol can be obtained by fermenting one tonne of sugar cane).

The future of ethanol power in Brazil is uncertain, for the government has not been following a consistent policy towards it. The sugar cane crop declined in the late 1980s, which made ethanol more expensive at a time when oil prices were relatively low. Consequently, the government issued a directive to the domestic motor industry to produce fewer ethanol-powered cars. However, the Gulf crisis of 1990-91 seemed to give alcohol fuel a new opportunity, and in late 1990 the Brazilian government expressed a willingness to support those alcohol producers which were prepared to increase their productivity. The problem was that sales of ethanol-powered cars had by then already collapsed because of the government's previous actions, and consumers could only be confused further by its latest policy switch.

Greater awareness of the potential value of alternative fuels has been emerging in Japan. In April 1991 the Japanese cabinet approved a white paper from the country's Environment Agency which called for increasing use of electric and methanol cars, along with a revision to the tax structure which would favour environmentally friendly forms of transport. The proposals had to be accepted by the Ministry of International Trade and Industry before legislation could be drafted, but the motor industry in Japan has been responding to the Agency's initiative by devoting greater attention to alternative fuels.

Electric vehicles

In Los Angeles, the preferred solution to the local pollution problem appears increasingly to be the electric vehicle. The City Council launched an Electric Vehicle Initiative in 1990, which aimed to study a possible transition from conventionally powered to electric vehicles. The idea was given further impetus by the exhaust emission ruling passed in late 1990 which, it has been forecast, will necessitate 70% (or over 6 million) of all vehicles operating in Southern California to be electrically powered by 2010.

Over 200 companies and agencies worldwide responded to the Electric Vehicle Initiative, and from these a Swedish-based company, CleanAir, was selected to produce an electric vehicle, the LA301. Developed by International Automotive Design (IAD) of Worthing, England, the new vehicle is a four-seater, three-door car, powered by a 43 kW electric motor and, when

The LA 301 petrol/electric hybrid car (photo Peter Cope)

additional mileage is necessary, a four-cylinder 650cc petrol engine. Because it has a petrol engine in addition to an electric motor, it cannot conform to the Zero Emission Vehicle category, but it is designed to meet Ultra Low Emission Vehicle (ULEV) standards, which are scheduled to come into play in the 1997 model year.

The LA301 should be on sale in the Los Angeles area in the spring of 1993, which means it will predate the entry into force of the ULEV ruling by more than four years. It will be made initially by IAD at a purpose-built plant in the UK, but larger-scale production facilities may be established later in the US, if demand merits such a step. Other cars will enter into competition with the LA301 as the decade progresses, and the electric car parc is bound to grow as the new Californian emissions regulations take effect. However, sales of electric cars will at first be restricted by their relatively high price and by the technical shortcomings of the available batteries. Demand will only increase significantly once new battery technology is available, allowing the problems of range and recharging time to be overcome. Then higher volume production will enable manufacturers to sell electric cars more cheaply.

It may be that only recourse to alternative fuels can provide a durable solution to the problem of local air pollution from motor vehicles for, as cities such as Los Angeles have found, the beneficial effects of reducing emissions from conventionally powered vehicles can be limited by increases in the vehicle population.

Keeping cars out

If there is to be a role for alternative fuel vehicles, it would appear initially to be for transport in urban areas, where the highest concentrations of air pollution are recorded. A number of cities around the world have in recent years resorted to emergency measures to combat high local levels of air pollution. In Europe, both Athens and Rome have been forced to adopt crude measures, based on registration plates, to limit the number of vehicles allowed to use the city on a particular day (although Rome has since abandoned this method), and in 1989 the Mayor of Milan exhorted people not to drive in the city. Florence has gone further than this, and prohibits cars from gaining access to the city between 0800 and 1830 hours, while in October 1991 the Greek government banned all private cars temporarily from the centre of Athens in order to counteract record levels of photo-chemical smog.

In other countries, such as Germany and Switzerland, notices have been introduced instructing motorists to switch off their engines while waiting at traffic lights, and the Netherlands has considered introducing tolls on urban roads at peak times in order to discourage unnecessary journeys. A road-pricing system has been tried experimentally in Oslo.

In Mexico City, officials introduced a policy in late 1989 banning hundreds of thousands of vehicles a day from the streets, despite opposition from within the motor industry, which preferred old cars (which are generally more polluting) to be singled out for prohibition. Car owners in the metropolitan area were banned from using their vehicles one day each week, in an effort to remove 400,000 cars a day from circulation. However, many residents circumvented the ban by purchasing a second car. Other cities have adopted emergency measures to combat the problem of traffic congestion which, although it constitutes a separate issue, is nonetheless linked with the question of vehicle pollution. One such example is provided by Osaka in Japan, which in April 1990 inaugurated a programme of 'No car days'. Road pricing has been tried in Singapore and Hong Kong. The debate on traffic in cities is assuming climactic proportions in many countries.

Legislating for fuel efficiency

Toxic emissions are not the whole story as far as car exhaust gases are concerned, because of the rapidly growing concern that output of carbon dioxide from motor vehicles is contributing to global warming. Here catalytic converters cannot play a positive role.

The only way to reduce carbon dioxide emissions is to use less fuel (this applies to industry and to domestic users of electricity as well as to motor vehicles), and there are growing calls in Europe for a 'carbon tax' which would penalize heavy emissions of carbon dioxide. In September 1991 the European Commission brought a carbon tax a little closer to reality by launching proposals for a new energy taxation system which would add $3 to a barrel of oil in 1993, rising to $10 in 2000. It also spoke of taxes which would take into account the amount of energy used by vehicles, and the possible introduction of EC-wide speed limits.

The 'environmental' taxation of cars can take various forms, and several methods are already applied throughout the world, even if purely ecological goals do not always provide the principal motive. In the US, for example, a 'gas guzzler' tax is levied on fuel-efficiency (a car's mpg rating), with the most profligate cars attracting the highest sums.

The gas guzzler tax was introduced in 1978, and is paid by the owner of a vehicle if its fuel efficiency is 5 mpg lower than the statutory average for that year (see below). Since 1986 any car with a fuel economy of less than 22.5 mpg has been subject to the tax, but the precise amount payable has varied between $500 and $3,850, depending on how much fuel the car consumes during the combined city/highway test cycle.

This tax was meant to steer commuters towards more fuel-efficient cars, but after the early 1980s its effect was limited by the relatively low price of petrol. However, the tax is still levied, and in November 1990 it was doubled by Congress, so that the new lower and upper limits are $1,000 and $7,700.

An alternative, or complementary, method to the one just described is to increase the tax on automotive fuel. The US government did this at the same time as it doubled the gas guzzler tax, but the increase was not great and petrol remained cheaper than in Europe. As mentioned above, raising the tax on fuel is also one of the instruments favoured by the European Commission to help it achieve a stabilization of carbon dioxide emissions.

In a number of European countries, cars are taxed according to their engine size. For example, the Italian government levies 38% VAT on new cars of over two litres, whereas the rate for cars with smaller engines is 19%. In France, the annual ownership tax on cars is levied at a rate which depends on the vehicle's CV rating, calculated on the basis of engine capacity and gearing. By 1992, no such differential was being applied in the UK, although there had been speculation for some time about the introduction of a progressive annual road tax linked to engine capacity, and about the higher taxation of company cars, which might steer people towards smaller-engined vehicles.

At present the price of petrol and diesel varies widely from one European Community country to another, largely as a result of the different level of tax that each country applies. Because of this disparity in taxation, petrol in Italy (the most expensive in the EC) costs around twice as much as it does in Greece (the cheapest in the EC). The UK falls somewhere in the middle; consumers have to pay more for petrol than they do in Germany, but less than they do in France or the Netherlands.

Diesel fuel is cheaper than petrol in every EC country, but here as well there is a wide gap in prices, with the Irish Republic turning out to be the most expensive and Greece once again the cheapest. Drivers of diesel-powered cars in the UK have to pay a relatively high price for their fuel, and there is little difference between the prices charged for petrol and diesel, whereas in some countries, such as France, Italy and Spain, diesel has a clear price advantage.

The European Commission has been trying for some time to harmonize the amount of excise duty that the governments of member states charge on petrol and diesel. Its original plan, presented in 1987, would have reduced petrol duty in Italy, the UK, Spain, Belgium, Denmark, the Irish Republic and the Netherlands, and have increased it in France, Greece, Germany and Luxembourg. Furthermore, the duty on diesel would have been reduced significantly in the UK, Italy, Denmark, the Irish Republic, France, Spain and Germany, and have been raised markedly in Greece, while remaining at roughly the same level as before in the Benelux countries.

Taken in isolation, the net result of these changes would have been to make fuel cheaper in many countries, including the UK, which would have represented a setback for the cause of fuel economy. This plan was subsequently dropped, and in early 1991 the Commission proposed a new system, setting more stringent targets for fuel duty. Under this system, the amount of duty charged on petrol would increase in most member states, with the exception of Italy and Portugal, and diesel duty would rise in France, Denmark, Germany, Belgium, the Netherlands, Luxemburg and Greece. A key consequence of these proposals was that diesel would enjoy a clear tax advantage over petrol (leaded or unleaded).

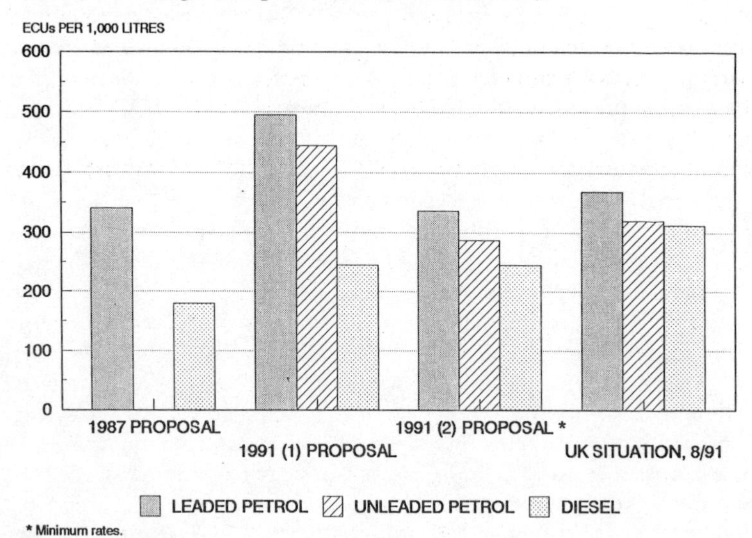

EC fuel duty proposals

This proposal was more ephemeral than the first, and by summer 1991 the Commission had backtracked. A new plan was presented, setting minimum tax rates per 1000 litres of fuel – ECU 337 for leaded petrol, ECU 287 for unleaded petrol, and ECU 245 for diesel. For petrol, the sums were far lower than those proposed earlier in the year, but there was little change in the case of diesel, so the preference given to diesel fuel was reduced.

The latest proposals were stalled when the environment ministers failed to concur on the whole package of tax reforms of which they formed a part, but they are likely to form the eventual basis of an agreement. Significantly, the room for manoeuvre which the minimum duty rates allow member states will eventually have to be relinquished, for the Commission's aim is still to achieve harmonization, possibly in 1996.

CAFE

Fiscal techniques are one possible method of steering car buyers towards more fuel-efficient cars, but in the US another method has been in existence since 1978. This is known as the Corporate Average Fuel Economy (CAFE), a statutory ruling under which manufacturers which sell cars in the US and produce more than 10,000 cars a year worldwide have to meet certain fuel economy levels. The standards are measured according to the average fuel economy of a manufacturer's range – individual models can fail to meet the specified level, but the average across the whole model range must be in conformity.

The system was not introduced out of any fears for the environment, but because the US government was concerned about conserving fuel in the aftermath of the major oil price rises of the mid-1970s. However, in the last couple of years the problem of global warming has been put forward as a justification for maintaining and even tightening fuel efficiency norms.

US CAFE standards, model years 1978-90 (city/highway USmpg rating)

1978	18.0
1979	19.0
1980	20.0
1981	22.0
1982	24.0
1983	26.0
1984	27.0
1985	27.5
1986	26.0
1987	26.0
1988	26.0
1989	26.5
1990	27.5

(Source: Motor Vehicle Manufacturers Association of the U.S., Inc.)

The first standard, set for the 1978 model year, was 18.0 USmpg, but the fuel economy requirements have been tightened up since then. The target was a 27.5 USmpg standard by 1985, and this level was in fact reached, but in 1986 it was reduced to 26.0 mpg as a result of lower petrol prices. It was kept at this level in 1987 and 1988, but rose to 27.5 mpg once more for the 1990 model year, where it has since remained.

CAFE standards will probably become more severe up to the turn of the century, as the perceived need to combat global warming grows. Recent attempts to tighten up the standards have been unsuccessful, but they provide an indication of the direction in which policy may move in the next few years. For example, the clean air bill which passed through the Senate in 1990 originally contained a carbon dioxide clause which, if it had been retained, would have resulted in a CAFE standard of 33 mpg by 1995 and 40 mpg by 2001. A separate bill introduced in the Senate by Richard Bryan (Nevada) called for each car manufacturer to achieve a 20% improvement in corporate average fuel economy by 1995 and a 40% improvement by 2001, using 1988 as the base year, but this too was defeated.

Up to now, European governments have generally preferred tax incentives to regulation as a means of favouring more fuel-efficient vehicles, but in Germany there is support for the establishment of average fuel economy standards from the opposition Social Democrats, and the European Parliament is in favour of similar measures, based on the introduction of a ceiling for carbon dioxide emissions from cars. Even a European car manufacturer – Peugeot – has expressed support for a CAFE-related system, although it would like to see average limits applied to all major pollutants and not simply to fuel efficiency. This kind of regulation may find increasing favour in the EC, especially if it is thought that tax incentives alone will not be sufficient to steer consumers towards more fuel-efficient cars.

Global warming moves to centre stage

Until recently the subjects of toxic emissions and global warming had been treated as separate problems; indeed, measures to combat the first do not necessarily help ameliorate the second, as the use of catalytic converters demonstrates. However, some policy-makers are beginning to treat individual environmental questions as part of a single, overarching problem. This is happening in West Germany. It has been suggested that cars be taxed on the basis of a system which takes into account all the exhaust gases that they emit, weighted so as to penalize those cars which emit a higher proportion of the most harmful gases. Just how carbon dioxide would be assessed in relation to nitrous oxides in any such system is uncertain, but this idea at least shows the direction in which environmental policy may move before long (cf. the Dutch Environmental Policy Plan, discussed in Chapter 9).

It is likely, nonetheless, that the problem of global warming will attain the centre stage and become the most fundamental of all the environmental challenges that the world has to face. Although it is somewhat ironic that this problem has emerged just as the West Europeans have decided to come to grips with the question of toxic emissions, the struggle to reduce and, where possible, eliminate emissions of the greenhouse gases may ultimately have greater repercussions on the motor industry than any efforts made to cut down the amount of toxic gases in the atmosphere.

These repercussions will depend on the extent to which the motor industry in the major producing regions – Europe, the US and Japan, plus their South American affiliates – is induced to abandon conventional forms of motive power in favour of alternatives which produce fewer or no greenhouse gases. Much will depend on the targets which are set by individual governments, or by supranational bodies such as the EC, for the stabilization or reduction of carbon dioxide emissions. Many governments are now concerned about carbon dioxide emissions, but not all are responding in the same way. Germany has stated its intention to reduce emissions by 25% between 1990 and 2000, while the UK is aiming for a mere stabilization at the 1990 level by 2005.

In October 1990 the EC member states agreed to achieve stabilization of carbon dioxide output by the end of the century, but individual countries will be able to proceed at a faster or a slower pace so long as the broad objective can be attained. The German target is highly ambitious, and will require a fundamental change in energy policy if it is to be fulfilled. Even the achievement of a stabilization is open to question. In the UK, for example, a doubling of road traffic between 1990 and 2025 has been forecast. This in itself must result in an increase in carbon dioxide emissions, unless there is a radical change in the type of fuel used.

The few alternative fuel policies that have so far emerged (see above) have been a response to localized pollution problems or economic imperatives rather than a sign of concern about the global level of carbon dioxide emissions. However, as evidence supporting global warming becomes more widely accepted, car manufacturers are taking more seriously the possibility of alternative fuel vehicles, and by the end of the century, depending on what the legislators decide, cars powered by other than petroleum-based fuel could be available on a significantly larger scale than they are at present.

4: *Making cars more efficient*

Many people both inside and outside the industry still think that by sticking on a catalyst and thus controlling toxic emissions, we have solved the problem and will drive happily ever after. This is only the beginning; the environmental impact of the motor car is far greater than its toxic emissions alone. Apart from the production process and the misery caused by the many traffic accidents involving cars, the product itself in its present form carries with it a number of inherent environmental problems. The key words here are: fuel efficiency, longer useful life and recyclability/renewability. Even on a smaller scale the way we maintain our cars still means that discarded oil, grease (a biodegradable grease only exists experimentally) and other fluids enter the environment. Even washing our cars uses vast amounts of water and can introduce detergents into the environment unnecessarily, although automatic car washes that recycle their water are increasingly used.

Greater fuel efficiency

In times of austerity, the car producers' abilities to develop, build and market fuel-efficient cars usually come to the fore and the immediate post-World War II period led to many small, aerodynamic and lightweight designs that achieved good fuel consumption. As prosperity increased and petrol became more affordable, fuel consumption became less important and cars grew bigger, squarer (less aerodynamic), fatter and heavier. Then the energy crisis of 1973/4 intervened and car producers once again made great strides in improving the fuel efficiency of cars. Over the last 15 years, the average car has become significantly more fuel-efficient.

Another impetus was given by the smaller energy squeeze of 1979, but since then, petrol has become cheaper and fuel efficiency has been virtually removed from the agenda. Unfortunately, whatever the present oil reserves may be, oil is a finite resource. The processes whereby oil is produced from decaying organisms over millions of years cannot keep up with the rate at which we extract it and whatever the current price of our petrol and diesel fuel, we should start thinking of the real environmental cost of burning away a finite resource and producing pollution in the process. We use oil for many other purposes, such as lubrication or the production of plastics, and we should consider whether it is not perhaps more sensible to use this finite resource in a more constructive manner, rather than letting it go up in toxic smoke.

Although the petrol engine is one of the more efficient solutions to the problem of powering a car, only some 28% of the energy in the petrol is available as power at the crankshaft (10-15% more in diesel engines). Of this remainder, around 10% is absorbed by the alternator and cooling fan, 6% by the gearbox, 4% by the differential, and 5% in the rest of the drivetrain. Rolling resistance absorbs another 18%, while air resistance takes a hefty 40%. Only 20% is then left to accelerate and go up hills.

There are several ways to make a car use less fuel. We can:

(1) change the engine by making it smaller and/or more efficient;
(2) change the shape and make it more aerodynamic;
(3) reduce weight/mass;
(4) change gearing;
(5) change operating conditions.

Apart from limiting the maximum speed (Mercedes-Benz and BMW now voluntarily limit the speed of their fastest cars to 250 kmh = 154 mph!), point number (5) is not completely under the control of the manufacturer, and will be discussed in Chapter 8, but the other points are and we will tackle these one by one. No car producer is going to implement such changes unless there are sound commercial reasons to do so. Another option is to introduce legislation forcing manufacturers to make more fuel efficient cars. This is what the US government did through its CAFE (Corporate Average Fuel Economy) legislation (see Chapter 3).

Engine efficiency

It may come as a surprise to many, that the modern petrol engine is actually remarkably efficient. Virtually any other source of energy or type of engine would need more room to produce the same amount of power, more complicated and larger fuel storage facilities, much tighter and more sophisticated engineering tolerances and production systems (see Chapter 7). This does not mean that the present offerings cannot be improved. Also, as emissions requirements become a more important factor, the equation may well change and make other systems more efficient.

The US government has classed a number of alternative fuels as 'clean'. It has used a rather limited criterion based on their contribution to tropospheric ozone pollution (smog) as found in a number of major cities. Unfortunately carbon dioxide emissions have largely been ignored and such alternative fuels as alcohol (ethanol, methanol) still produce high levels of carbon dioxide, as do all hydrocarbon-based fuels. Only if they are produced from crops – which absorb as much carbon dioxide as is released when they burn – can we avoid a net increase in carbon dioxide from such fuels, but feeding a growing world population is really a higher priority than powering our cars. The real answers to the pollution question seem to be hydrogen or electricity. Both of these have to be produced first and at present the wide network of non-hydrocarbon production systems for these energy sources (e.g. wind, solar, tidal, hydro) is not always available. Besides, both systems require heavy and complex storage systems.

For the time being, most of our cars are therefore likely to be powered by petrol or diesel engines. These can take many different shapes and some recent designs, such as improved two-stroke, Orbital and others, may well reach the production stage. Most of these combine even greater efficiency with greater simplicity and lower weight.

Even the present designs are still being improved; multi-valve heads, turbo-charging, fuel injection, engine management systems all enable us to get more power from the same amount of fuel. A logical step would therefore be a radical reduction in engine size towards small engines of much greater sophistication. This would combine lower toxic emissions in absolute terms with lower carbon dioxide emissions and lower fuel consumption.

Engine designers recognize a point of maximum efficiency with cylinders of around 500cc; i.e. a 4 cylinder 2 litre engine, a 6 cylinder 3 litre engine, or even a 2 cylinder 1 litre engine. This means that a 4 cylinder engine of 500cc for example is inherently less efficient than a 4 cylinder engine of 2 litres capacity. Although the smaller engine does not use the expected 25% of fuel that the larger engine uses, it will, nonetheless, always use less and such a small engine could therefore still be justified. What can be achieved when mating such a tiny engine with modern technology is illustrated by the Mitsubishi Minica Dangan, only available in Japan, which has the following specification: it has 3 cylinders with 5 valves per cylinder, a capacity of 550cc, fuel injection, a turbocharger and an intercooler, it gives 63 bhp at 7500 rpm, and its top speed is limited to 87 mph (140 kph).

We can compare this with the Metro City, which achieves the same 63bhp power output with 4 cylinders, 2 valves per cylinder and 1275cc. In Japan, legislation favours such tiny cars, although for the manufacturer they are hardly cheaper to produce than larger models.

Streamlining and aerodynamics

Close links with aircraft production in World War I first made the motor industry aware of the benefits of streamlining. Early examples of streamlined cars were limited to the racetrack, where Bugatti, Chenard-Walcker, Voisin and others amply demonstrated their advantages. The initiative was then taken over by specialist aerodynamicists like Rumpler and Kamm in Germany, Burney in the UK, Andreau in France and the Hungarian-born Jaray in Switzerland. At the same time the less scientific stylists made streamlining fashionable. Both aerodynamics and streamlining signified speed and racing cars began to benefit from the new science during the 1930s. After the war, the frugal climate led to the need for small, cheap, fuel-efficient cars and aerodynamics spread to the small production car, as exemplified by the postwar Saab 92/93, the Panhard Dyna, Messerschmitt and several small sports cars.

During the 1960s slab-sided squared-off cars abandoned aerodynamics to a large extent, although Saab and Citroën in particular persisted with their designs and so did the sports car producers or their designers; Italian coachbuilder Zagato, for example, established a reputation for light aerodynamic bodies on production chassis. It was once again the energy crisis of 1973/4 that put aerodynamics back on the car producers' shopping lists and several milestone designs such as the Citroën CX of 1974 and Audi 100 of 1983 helped bring down the average air resistance of the modern car to a Cd of around .32 (Cd is the coefficient of aerodynamic drag; a theoretical upright square has a Cd of 1.0). Clearly, improvements are still possible and some experimental designs have shown just what is feasible even today; the Citroën Activa prototype, for example, has a Cd of 0.25, while VW's experimental ARVW even manages 0.15! (Figures of around 0.19 had already been achieved on prototypes in the 1930s).

Weight the enemy

This phrase was commonly used by French racing car designers in the 1920s (le poids: l'ennemi) and probably originated with Ettore Bugatti. Car makers realized fairly early on that by making a car lighter you could make it go faster, or at least accelerate faster, often crucial in competition. They also discovered that a reduction in weight led to a reduction in fuel consumption and this was useful not only in competition, but also in the new cheap popular cars developed especially from the 1920s onwards.

New techniques and new materials were introduced to reduce the weight of cars. On the more expensive and sporty cars, aluminium bodywork was used, while cheaper cars often featured fabric bodies, pioneered by Weymann. From the 1950s onwards, the newly developed plastics were increasingly used, especially for low volume sports cars (Chevrolet Corvette, Lotus, TVR) or small cars (Reliant, Bond, Trabant).

The lighter the car, the less weight needs to be accelerated from standstill to cruising speed each time a car is driven, and the less fuel is therefore used. It is a sad fact, however, that many cars put on weight over their lifetime. This tendency has become more marked over the last twenty years or so and even the cheapest, simplest cars have been affected by this middle age spread as the table overleaf shows.

Some of these grew larger engines, most became more comfortable, received better instruments, new safety equipment, etc. and the net result is an increase in weight. The Mini is the odd man out; its weight, although fluctuating with changes in basic specification, is now remarkably similar to the weight of the original car in 1960.

Weight increase of small cars 1960-1990 (in kg)

	1960	*1965*	*1970*	*1975*	*1980*	*1985*	*1990*
VW Beetle	731	780	800	820	780	780	820
Citroën 2 CV	499	510	540	560	560	600	600
Renault 4	575	575	600	695	695	665	720
but:							
Mini	634	634	600	615	620	630	630

The picture is a lot worse in the larger cars. Here, the proliferation of gadgets that plague the modern car have exacted a heavy toll; electric windows, sunroofs, seats and mirrors; audio equipment, spoilers, larger glass area and increasing body size all add weight. Other factors like improved passenger safety and pollution control equipment can be justified, though they too add weight. Here are the weights of gadgets on the BMW 735i: electric sunroof 17kg, central locking 3kg, electric seat adjustment 10kg, electric mirror 1.2kg, climate control 25kg, audio equipment 10-17kg, anti-lock brakes 10kg, headlamp wash/wipe 10kg, self-levelling suspension 13kg, rear sunblinds 2.7kg, towbar 26kg.

Some manufacturers have actually reduced the weight of the basic body structure significantly. Mercedes-Benz for example has managed to reduce the weight of the basic car with each successive generation of its intermediate models (current 200-300); however, because the market requires more and more equipment, most of this weight reduction is wasted. Competitors BMW and Audi have been less successful, as each new model was heavier than its predecessor (see table below).

Weight increase of medium to large cars (in kg)

Model	*1967*	*1977*	*1987*
Audi 80/90	880	933	1070
BMW 1600/316	936	1050	1110
Ford Zephyr/Granada	1240	1310	1330
Fiat 125/132/Croma	1050	1180	1230
Porsche 911	1090	1160	1270
VW Golf		860	930
but:			
Mercedes 200	1375	1390	1370

(Adapted from Auto Motor und Sport)

As new items for the enhancement of our comfort, convenience and safety become available, we may well see further increases in weight. Greater use of lighter materials may to some extent offset this. Plastics are increasingly used and so is aluminium. But to achieve real improvements, a drastic reduction in weight is needed.

After the oil crisis of the 1970s, American buyers changed their buying behaviour from the traditional US product to smaller and more fuel-efficient cars from Europe and Japan. US producers responded to this new threat by a radical programme of downsizing. Initially this process spawned some strange hybrids like the compact but still heavy AMC Pacer, but over the past few years, US cars have become lighter, more fuel-efficient and much better into the bargain. This episode shows what can be done under consumer pressure; a lesson for the future.

In Europe and Japan, meanwhile, consumer preference has favoured ever larger and heavier cars. Gains made in fuel efficiency are partly undone by increases in weight, size and performance. Some manufacturers recognize the problem and radical design proposals such as the lightweight high performance Matra M25 prototype (720 kgs, 200 hp) and the production cars like the Citroën AX show the way.

A Dutch team at the technical university of Delft has been proposing the lightweight car concept. This would involve a radical reduction in weight and size. The team reckon that following existing trends they would expect to see between 1990 and the year 2000 a weight reduction of 5%, possibly a higher operational efficiency of the internal combustion engine, and a marginal increase of Cd to an average fleet figure of 0.30.

This is not enough and developments are too slow to make any real environmental difference. The team concludes, therefore that 'the key to lower fuel consumption of vehicles and fleet is introducing the very light car as soon as possible'. The concept they have developed would involve two scenarios, one allowing for a maximum weight of 800 kgs, the other for a maximum weight of 600 kgs. The latter especially could lead to a drastic reduction in carbon dioxide emissions. These cars would be less comfortable than most of today's models and would be used for shorter distances; commuting would be reduced and the car would become a less versatile means of private transport, while longer distances (beyond 50 miles or so) would be covered by an efficient public transport system. The team also suggest a strong increase in fuel taxes as an incentive to move in this direction, as well as US-style maximum fuel consumption standards (CAFE).

Many economical lightweight cars were developed in the 1940s and 1950s, such as the Citroën 2CV and Panhard Dyna, but none more so perhaps than the French Mathis VEL.333, the Voiture Economique et Légère, which featured 3 wheels, 3 seats and had a fuel consumption of 3 litres/100 kms (around 92 mpg). This was achieved through a low weight of 612 kgs (aluminium body), a small 2 cylinder engine of 707cc and exceptionally good aerodynamics (Cd of only 0.20!).

Safety

Between 1977 and 1987, 57,700 people were killed on Britain's roads, while a staggering 3.1 million people were injured. However, public transport accidents account for only 1% of all transport fatalities. Despite the publicity surrounding train and coach crashes, cars kill many times more people, being responsible not only for driver and passenger casualties, but also for most motor cyclist, pedestrian and cyclist deaths. In fact, the UK compares particularly badly in terms of casualty rates for these last two groups, with pedestrian accident rates more than three times higher than in Sweden or the Netherlands, while cycle casualty rates are ten times that of these two countries, according to a study by Tight and Carsen in 1989 (see *Vital Travel Statistics*, EERU).

This illustrates the point that apart from inadequate pedestrian and cycle provisions in Britain, it is very often the case that the heavier the vehicle, the safer it is. This militates against reducing the weight of cars. Some of the weight added in successive generations of cars is attributable to increased driver and passenger safety and to many buyers this may be a more important criterion than the environmental acceptability of their car. A reduction in weight of all cars and a separation of different modes of transport is the only answer. In fact, automotive engineers have been able to redesign the basic body shell so that it has shed weight as well as increasing crash-resistance over the past two decades.

As well as the human cost, there are also financial implications of these disastrous statistics. Accident and emergency systems have to be in place, scarce hospital beds are taken up, and there are the spiralling costs of hold-ups and congestion that result from an accident.

Gearing

Another important aspect of fuel-efficient design concerns the gearing of the car. The widespread introduction of 5-speed gearboxes has made a significant contribution to the greater fuel efficiency of today's cars compared with 15 years ago. Some cars are even being fitted with 6-speed gearboxes for higher speed and lower consumption, while automatics are also following the trend. Most automatics are now 4 instead of 3-speed and some 5-speed models are beginning to appear.

An alternative is the continuously variable transmission or CVT. This system does not use 4, 5 or even 6 fixed ratios between the engine and the wheels (i.e. the gears in your gearbox), but provides an unlimited range of ratios, so that the engine can always operate at its ideal performance level, whatever the speed or road conditions. As a result, a smaller, more fuel-efficient engine can be used, without a noticeable reduction in performance. Some manually operated CVTs were used on early cars, especially in Switzerland (Weber 1902, Turicum 1904) and France (Fouillaron).

The first automatic CVT to be put into production on a car was the Variomatic, the main feature of the Dutch DAF car introduced in 1958. This system uses two sets of variable diameter pulleys. The first set is driven by the engine via an automatic clutch and their diameter can be varied by the engine vacuum. These drive the second set of pulleys – whose diameter is changed by road speed and rolling resistance – via rubber belts, and from there the wheels are driven. The system provides built-in traction control

Transport accidents in the UK

	1977		1981		1987		1977-87 Percent change	
	Killed	*All severities*	*Killed*	*All severities*	*Killed*	*All severities*	*Killed*	*All severities*
Child pedestrians	440	30,374	318	23,722	245	19,934	−44	−34
Child cyclist	95	9,705	86	9,358	68	7,934	−28	−18
Adult pedestrians	1,869	38,977	1,552	35,299	1,454	36,587	−22	−6
Adult cyclists	206	13,392	224	15,570	212	18,479	+3	+38
Motorcyclists	1,182	71,689	1,131	69,129	723	45,801	−38	−36
Car drivers	1,429	82,485	1,346	81,079	1,327	92,010	−7	+11
Car passengers	1,012	69,025	941	65,238	879	67,458	−13	−2
Light goods vehicle drivers & pass,.	152	11,369	141	9,111	111	8,842	−27	−22
Heavy goods vehicle drivers & pass.	109	4,491	67	3,044	75	3,487	−31	−22
Bus/coach pass. & drivers	64	12,375	20	9,886	15	9,088	−76	−27
All road	6,558	343,882	5,826	321,436	5,109	309,620	−22	−10
All rail	55	2,173	52	2,437	47	2,774	−15	+28
All air	31	47	43	75	65	48	n/a	n/a

(*Source: EERU, Vital Travel Statistics*).

and the earlier versions also worked as a limited-slip differential; both of these features are normally available as expensive options only on upmarket cars, but the DAF had them for a bargain basement price. The system was last offered in some countries on the Volvo 340, but an improved, more compact, version has been developed by the Dutch firm Van Doorne's Transmissie, using a steel belt, and this is available for Ford's Fiesta, the Fiat Uno, Lancia Y10 and Subaru Justy. The advantages of this system are now widely recognized and we are likely to see more cars fitted with it, especially in the smaller segments.

Planned obsolescence; a terminal disease?

On recent estimates, about 40% of the energy a car uses is consumed during its manufacture. If we are going to save our – expensively extracted – finite resources, not only should we make cars more fuel-efficient, but we should also make them last longer. If they last longer, we need to replace them less often, and we therefore need to make fewer (but better – and possibly more expensive) cars and save on the vast amounts of energy they cost to produce.

The Japanese have adopted a four-year product cycle for most of their more popular models; roughly every four years we get a new Toyota Starlet, Honda Civic, Nissan Bluebird, etc. This allows the incorporation of the latest Japanese technology and gadgetry in what can be presented as an all-new model. This method seems very similar to the annual facelift practised by the US manufacturers in the 1950s and 1960s. The problem is that the earlier model is immediately regarded as obsolete by many consumers and major sheetmetal and other changes may lead to the rapid phasing out of spares for older models. In Japan itself, cars are also discarded long before their useful life has come to an end. Many such cars are exported to poorer southeast Asian markets. An exception is Isuzu, which reconditions older examples of its own make and markets them via the Musha network; the number of older examples of the make is indeed noticeable in Japan.

This attitude is now being imposed on their European competitors, many of whom feel that the regular renewal of the model range gives the Japanese a competitive advantage. The Europeans have traditionally kept a model in production for a relatively long time. During this long production cycle, improvements are constantly being implemented, but the basic structure remains unchanged. As a result, the consumer feels that his car is not yet obsolete and parts availability is rarely a problem. Extreme examples of this approach are the Citroën 2CV, first developed in the 1930s, launched in 1948 and still produced in Portugal until July 1990, albeit in a much improved form. More mainstream products also achieve much longer replacement cycles than those in Japan; Ford's original Fiesta was replaced after more than 12 years.

The European specialist producers have an even better record in this respect. In fact, longevity has always been regarded as a sign of quality in Europe and the prestige image of companies like Mercedes-Benz and Volvo is largely based on the fact that their products not only have a very good average life expectancy (over 20 years for a Volvo), but also on the fact that the same models stay in production for a very long time. The consumer feels that they must have got it right in the first place! Each generation of the Mercedes S-Klasse, one of the world's most prestigious cars, stays in production for more than 10 years. Jaguar produced the same basic model for 17 years, while the Volvo 200 series, first introduced in 1974 and still a strong seller, dates back to the original 140, launched in 1966/7. Saab too believes in long cycles; the 900 was derived from the 99, first introduced in 1966/7.

In some countries, and West Germany is one example, very tough periodic testing removes many cars prematurely from the roads. In the case of Germany – like Japan – quite a few of these cars are exported; to Eastern Europe and Greece. Nevertheless, the average life expectancy of cars in West Germany and other industrialized countries has been getting longer. In the Netherlands, the average car in 1989 lived for 10.5 years (from 8.2 years in 1978), in West Germany, France and the UK for 11 to 12 years, in rust-free Italy and Spain 15 years, while the US achieved an average 14 years of use out of its cars.

German manufacturers have made real efforts to extend the life expectancy of their cars still further as is illustrated by the long life of a Mercedes and Audi's well publicized decision to use galvanized steel for the entire body-shell, something which Porsche has in fact been doing for some time. Worldwide, engine life has also increased markedly over the past three decades, thus making older cars more attractive to run.

Porsche has also pursued this alternative view in its experimental 'long-life automobile' research project. An interim report on the project concludes that:

> Long-life automobiles could . . . be produced in volume and reduce the drain on our energy and raw material resources . . . In cost-benefit terms the desirable lifespan of 20 years or 300,000 km (185,000 miles) distance covered could be reached if the long-life automobile were to be constructed largely of aluminium. This would also permit the maximum saving in energy costs and raw materials.

'Classic' cars

The growth of the classic car movement is a relatively recent phenomenon. Although investors have now jumped on this bandwagon, most enthusiasts just like their old cars and in the process they show that it is perfectly possible to keep a car going and looking good for 20, 30, 40 or even 50 or more years. Although few classics conform to present or future emissions standards, they provide usable alternatives that are only a limited threat to the Earth's resources compared to the production of a new car. Organizations like the Morris Minor Centre have shown that it is possible to make money by keeping old cars going *ad infinitum*, adding some improvements (better brakes, more powerful engines, better rust protection, etc.) along the way. Other companies have followed by rebuilding classic cars in a kind of series production, turning out products that are often better and usually more durable than the original when new. Thus one can now choose various series-rebuilt models from a very expensive Vicarage Jaguar Mk2 or E Type, or a Triumph Stag from Tudor Classics, to a Citroën DS rebuilt in France and sold in the UK by the Morton Stockwell Group. Other companies, like Lotus and Rolls-Royce, still regularly produce spares for long obsolete models.

This is a trend that should not be ignored and should be considered a realistic alternative to the purchase of a new car. The above are all enthusiasts' cars, for which there exists considerable demand, but any solidly built and relatively simple car could qualify as a long-life car within this context. Apart from the Morris Minor, such cars as Volkswagen Beetles, old Volvos, Saabs and Mercedes all qualify if properly looked after by the first few owners.

Renewable resources and recycling the car

As well as extending the life of the car, we should make sure that we reduce the depletion of scarce resources by using the materials again when our vehicle has really reached the end of its useful life. Recycling is becoming a high priority, but has in fact been around for decades in a limited form. The scrapyard is the simplest type of car recycling centre. People who drive older or classic cars can often find usable parts in a scrapyard. Items like batteries normally go for recycling straightaway (new batteries contain 90% recycled lead), while the steel shell once it is stripped can easily be recycled. Even waste paper is often used by manufacturers to produce insulating materials. Glass is also almost fully recyclable – although this is rarely done today – while modern electronic units can often be reused in their entirety. In most developed countries around 80% of all parts that make up a car are presently recycled, and this figure could be increased.

Plastics are a growing problem, however, because they are increasingly used on cars. In fact, between 1965 and 1990, the steel content in cars dropped from 76% to around 66%, while the plastics content rose from 2% to 10%, and will have reached around 15% by 1995. Although plastics have advantages in terms of lightness and mouldability for good aerodynamics, they have been presenting problems. Most plastics cannot be recycled and cause problems when mixed with recyclable metals. There are three major types of plastics:

(1) thermosetting; around 80% of plastics and are non-recyclable (e.g. GRP);
(2) elastomers; non-recyclable;
(3) thermoplastics; recyclable up to 3 or 4 times.

Plastics manufacturers realize the problem exists and are trying to take action. Recyclability is being studied, recyclable plastics are being developed and special codes are also being introduced, which help identify which materials can be recycled and by which process. General Electric Plastics has in fact set up a pilot scheme under which it buys back around 300 tonnes of plastics through an arrangement with scrapyards in the Munich area.

Some car manufacturers too have taken the initiative and started recycling programmes. Companies like Volvo, Mercedes-Benz, Saab, Volkswagen, BMW and others are actively pursuing a policy whereby they regard themselves as responsible for the car's environmental impact during its entire life. Volvo is now designing cars taking into account their recyclability and other environmental implications (short of stopping car production), while BMW is building a special recycling plant at Wackersdorf in Bavaria. Volkswagen too is developing the technology of dismantling cars in large volumes and the company has proposed introducing legislation which would require the trading in of a car to be dismantled as a condition of buying a new one. A spokesman from the Dutch company Volvo Car BV suggests a deposit of around £150 per car to be paid on delivery. These deposits would be put into a fund to build up a proper recycling infrastructure and when the last owner scraps the car, he/she gets the deposit back. Such a system would obviously have to be set up by government, but it should yield enough money (invested for the lifespan of the car) to do the job. The UK government has suggested a similar scheme for tyres only.

In France, PSA is setting up a recycling facility near Lyon which is aimed at producing 'zero discharge'; any waste produced is to be put to some use. Together with its compatriot Renault, the firm has also developed one of the first coding systems for plastics to facilitate recycling and this has already been applied on the Citroën ZX. Together with Fiat, PSA has set up a

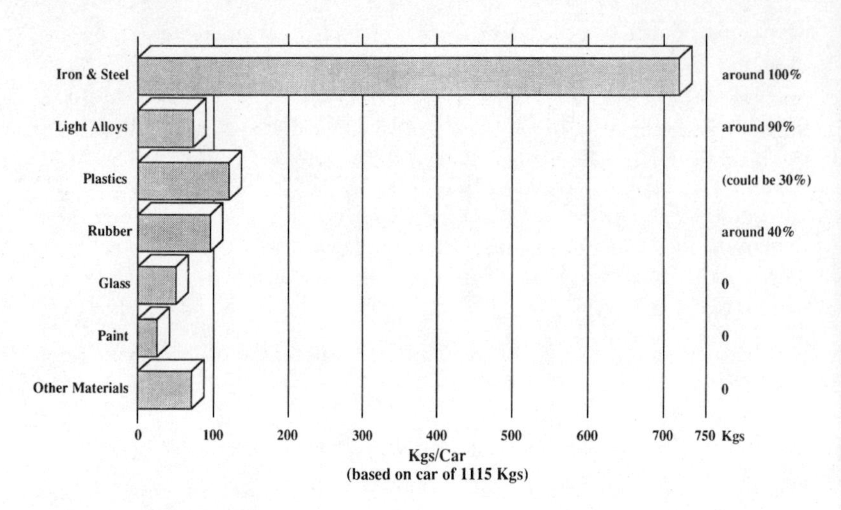

Re-use of car scrap
(Source: Autovisie 23.6.90).

research programme into plastics recycling under the title RECAP (RECycling of Automobile Plastics), which is linked with the EC Eureka research programme.

If no solution is found or the infrastructure for recycling plastics is lacking, aluminium may well be a more sensible alternative. Metals are much easier to recycle on the whole and aluminium is particularly attractive. Although it is very energy-intensive to produce from ore, at 15 to 18 kWh per kg, only 5% of this energy is needed to recycle aluminium. Not surprisingly, in countries like Germany or the Netherlands, 90% of all aluminium from scrapped cars is already recycled today.

A new type of recycling has come in the wake of emissions control. Catalytic converters use small amounts of the extremely rare metals rhodium (3 grammes in each catalyst) and platinum (1 gramme in each catalyst). With the growing number of catalysts used, the world's resources of these metals would rapidly run out, so several manufacturers now pay people for each catalyst returned to them and these valuable metals are recovered. Cheaper alternatives are also being sought, partly to reduce our reliance on countries like South Africa, one of the few suppliers, while closer links with the former USSR are beginning to improve supplies from there.

If we cannot recycle, we can always renew, but few renewable resources are immediately useful for building cars (see Chapter 7). The most obvious renewable construction material is wood and wooden cars have always existed, but are unsuitable for volume production.

After a car has lived its present average useful life of 10 years or so, we have three main options: to recycle its component materials, to retain it (perhaps to become a 'classic', or just as an 'old banger'), or to rebuilt it (as a series, or as a one-off). All these options are viable today and are options open to the individual driver. Fortunately, manufacturers themselves are beginning to take at least the recycling issue seriously, and we should see significant progress on this front throughout the 1990s.

5: *Car production: past, present and future*

In this chapter we will look into how a car is built, why it is built the way it is, what the implications are of this and what possible alternatives are available to us.

History

When the car industry started in Germany in the 1880s, the major innovation that led to the motorized vehicles of Daimler, Benz and others involved engine technology. A light, relatively fast-running engine made it suitable for fitting to a simple chassis. Benz's first car, his Patentmotorwagen, was a light 3-wheeler, essentially designed for the purpose and using existing bicycle technology. Daimler, on the other hand, first fitted his engine to a motorbike and later to a horseless carriage in the true sense. As the car developed, the first people who were geared up to produce them in reasonable numbers were obviously the carriage builders and bicycle makers. The first group were used to certain production methods and these were not changed significantly for the production of cars. Wood was the basic material of horsedrawn vehicles, often with a steel or steel-reinforced frame, and these techniques were transferred to the motor car. Bicycle producers, on the other hand, had learned to use tubular steel frames and generally lighter constructions, so we see in the early cars elements from both these traditions.

In those days, each car was built by hand on a separate chassis, and fitted with a wooden or timber-frame body. The first to abandon the horseless carriage idea and design a car from first principles was probably Lanchester. Lanchester's cars looked like nothing else available at the time. His early cars, introduced at the turn of the century, were low-slung and had the engine in the middle, but they still retained the chassis and wooden or wood-frame body. There was the French Latil of 1899, with a unitary-construction body, but few people saw the significance of this eccentric front wheel drive car.

By 1905, car production had begun to leave the small workshop and purpose-built car plants geared to volume production were built; by 1905-6, the Scottish firm of Argyll with its purpose-built factory in Alexandria near Loch Lomond produced more cars than anybody else in Europe (the building still exists, though now derelict), while across the Atlantic, some firms reached even higher numbers. Henry Ford made car production reach a greater order of magnitude by adapting the production line to the production of a simple car, although he still retained the same old method of construction. If it had been up to him, we would all still be driving black Model T Fords!

Some changes had already begun to take place, however. Aluminium was now often used to cover the wooden frame and by the 1920s this had almost become obligatory for all luxury and sports cars. Developments in the First World War also allowed aluminium drivetrain components (engine, transmission) to be added to this. Wooden chassis were now rare as steel had become the standard material, the first all-steel chassis being produced by Maybach in 1901. Some companies experimented with unitary construction bodywork, lighter and more rigid than the conventional method. Lancia moved in that direction with its Lambda of the 1920s, which used a light 3-dimensional frame to which the body panels were attached. Voisin went one step further and used a true integral or 'monocoque' construction for his 1923 grand prix car, which was made up of a reinforced aluminium tub, inspired by an aircraft fuselage, but this construction did not reach true series production.

It was generally recognized that the conventional steel chassis with separate body was too heavy and various refinements were being introduced. Ledwinka in Czechoslovakia developed the backbone chassis – one large central tube to which the major components were attached. Ledwinka first used it on a Tatra car, but others followed and the principle is still used by Lotus and TVR. Once pressed steel was introduced, some companies adopted a pressed steel platform to which the body was bolted; this system still featured in the VW Beetle and the Citroën 2CV.

Budd and the all-steel body

There were a number of major problems to be overcome. Existing techniques could not produce bodies in the high volumes now required. Lighter high-performance engines and lighter cheaper cars required lighter, cheaper bodywork, while the new fashion of streamlining and aerodynamics required more complex shapes, which were labour-intensive and expensive to produce. The solution to all these problems was provided by an American, Edward Gowen Budd, who developed the all-steel body. Apart from solving these problems, an advantage of the system was that the paint could be baked onto the bodies without fear of wood catching fire. This reduced the painting and drying process from several weeks to just one day!

Budd set up his own company in Philadelphia in 1912 and started the development of the all-steel body. In 1914, enough expertise had finally been accumulated for the first patent to be awarded to the company for an all-steel, all-welded car body and the Dodge brothers became the first major client. The first European to realize the potential of the Budd system was André Citroën. Gradually others followed and by the 1950s, only some traditional quality cars (Rolls-Royce, Delage) and sports cars still used a separate chassis.

The supply industry

Another major change in car production since the early days is the development of the supply industry. Around the turn of the century, most producers still had to make virtually every single part themselves. Even such items as carburettors had to be produced in-house before there were enough suppliers to produce these for everyone. Some firms continued to make even such minor items as nuts and bolts themselves to control their quality. However, suppliers did exist right from the start. Coachbuilders have already been mentioned and others developed alongside. As the new industry grew so did the number of specialist suppliers. Car manufacturers themselves often became suppliers to their competitors when they had developed a particular expertise. Thus Daimler and De Dion-Bouton both became important suppliers of car engines.

The growth in importance of the component supply industry continues today. Development costs become ever higher as cars and their components become more complex and sophisticated. Another problem is that each car producer is now expected to produce quality and excellence in every aspect of its cars. In the past, producers could produce a very good chassis with a rather uninspiring engine or a marvellous engine in a car with dodgy handling. Producers each had their particular area of ability. Nowadays this is not enough and firms that lack the know-how in particular areas go to outsiders for help. Alternatively, some suppliers have developed such expertise in a specific field that most firms buy the relevant components from them without ever considering doing it themselves. Thus many producers go to Bosch for fuel injection equipment, to Teves, Girling and others for brake parts, to Solex, Dellorto, Weber, SU and their competitors for carburettors and Borg-Warner, ZF or Getrag for transmissions.

Car production today

The main development of the production line, a strong supply industry and welded steel body construction have been instrumental in creating the modern car industry. A modern car plant is therefore basically an assembly plant with components and materials coming in from many different sources to be made into a car. The main in-house aspect of this process is the production of the bodyshell using developments of the principles pioneered by Budd and others. Production lines of various types still dominate in most assembly plants. The Budd-type unitary welded steel body forms the basis of all modern mass-production cars. The mass-produced car starts life as a roll of sheet metal, which then goes through the following processes in the car plant:

(1) *Blanking*: this is the process whereby the required shapes are cut out of the roll of sheet steel; door shapes, roof shapes, bonnet shapes, etc., although they are not always easily recognized as such.

(2) *Pressing or stamping*: large presses are used to give the blanks three-dimensional shapes. Each blank usually has to be pressed several times by different presses and each time the shape is changed. Many modern press shops use transfer presses which have different dies (the tools that actually shape the steel) moving at the same time; the steel blanks are moved to the next position each time the press moves up and when it comes down a new shape is pressed into it by the next die. Out the other end comes a door, a roof, a front wing, etc. More and more manufacturers also include moulding equipment in the press shop for plastic bumpers, petrol tanks, etc.

(3) *Body shop*: here the pressings are welded into a body shape. First the floorpan or platform is welded together from the different pieces. Then the body-sides are usually added and finally the roof. At each point, the first welds are usually put on in a fixed jig to make sure the dimensions are precise. Subsequent welds reinforce the body and are usually carried out by robots. Finally the opening parts are fitted: doors, boot lid, bonnet, and sometimes wings are welded or bolted on at this stage. From very thin and flexible steel, a rigid bodyshell is constructed.

(4) *Paintshop*: the welded body is now cleaned (degreased), etched, primed and painted. Rustproofing often takes place at this point and underbody protection is added. The paintshop is an unpleasant environment for people and robots are increasingly used to spray paint onto the bodies. After painting, the doors are often removed nowadays to avoid damage in final assembly and allow more room around the assembly line. This 'doors off' system was pioneered by the Japanese, who also argue that a car with open doors is much wider than a car without doors, thus wasting valuable floorspace on wider assembly lines. The doors have to be fitted to the same car again in final assembly.

(5) *Assembly*: this can be divided into two separate stages: before the engine and gearbox are added, and after they are added. These stages are called by different names: trim and final assembly, pre-assembly and final assembly, etc. The car moves along the assembly line, or some plants – such as the Dutch Volvo plant – now even use automatic guided vehicles on which the assembly workers travel through the factory with the car.

(6) *Trim*: here the wiring, brake pipes, fuel lines, interior trim (usually), glass, seats and fascia panel are fitted. Most of these processes involve manual labour, even in the most automated plants.

(7) *Marriage*: this is where the engine and gearbox are fitted to the body.

(8) *Final assembly*: during this process the car is finished: suspension and wheels are fitted and the car is fitted with doors, grills, rubbing strips etc., filled up with oil, water, petrol and other fluids and is finally tested.

Environmental implications

Iron

Several aspects of this process can be harmful to the environment. Most cars are still made of steel and steel is made from iron, which is smelted from iron ore, a rocky formation dug up from the earth. Each of these processes uses energy – normally from finite resources of fossil fuels – and produces pollution of one sort or another. Besides, although iron is one of the world's more common elements, ultimately it must be finite. Nevertheless, in view of the lost cost of steel and the very high capital investments already in place for the production of steel car bodies, most of our cars will probably continue to be made from steel for some time to come.

Other metals

Other metals are also used in a car; copper for the wiring, aluminium and various alloys, chromium for plating, nickel, lead (batteries), zinc etc. Aluminium production from bauxite (aluminium ore) is particularly energy-intensive (see Chapter 4) and many aluminium smelters are located in mountainous areas to use relatively cheap hydro-electricity to provide the vast energy requirements needed. Of course this form of energy is renewable, though it often disturbs the environment in other ways (such as damming of rivers and flooding of valleys). Zinc plating, although a major contributor to extending the useful life of modern steel car bodies, is also a major source of pollution in many areas.

Plastics

Raw materials for most plastics are fossil fuels such as oil, gas or coal. These have to be extracted as for fuel use and transported to the chemical plants which turn them into plastics. The chemical industry is known worldwide as a major polluter and attempts to clean up its act have come relatively recently, if at all. With the opening up of Eastern Europe several horror stories involving chemical plants are emerging, while other companies have set up plants in Third World countries where legislation or its enforcement is less strict than in the developed world.

Paint

The paintshop typically represents between 30% and 50% of the total cost of an assembly plant. Furthermore, painting can account for as much as 20% of the total cost of producing a car. Paint technology has changed drastically over the past 10 to 15 years as the chemical industry tries to meet the needs of ever more critical consumers – the paint is the first thing you notice about a car after all – and the need for longer lasting cars, as well as the increasing environmental pressures. The days of Henry Ford, who only offered black cars (black was the fastest drying colour in those days) are long gone. However, the paintshop is the single biggest source of pollution from any car assembly plant.

Paint is a product of the chemical industry. Pollution from paintshops at car assembly plants has always been a feature of car production; you can often smell a car plant when you pass it and what you smell is the solvents used in paint. These solvents are hydrocarbons classed as volatile organic compounds (VOC). VOCs mix with nitrous oxides to form low-level ozone and they are increasingly the subject of environmental concern. In the motor industry state of Michigan, for example, some 23.5% of all VOC emissions originate from car paint; other sources are petrol fumes from fuel stations and leaking car fuel systems, as well as drycleaning solvents.

The paint has to be heavily diluted with solvents in order for it to be pumped through pipes and sprayed through the spraygun nozzles. Certain colours contain greater quantities of harmful materials than others. Metallic paints, for example, require a greater VOC content than solid colours, partly due to the fact that they require an extra final clear coat. This has already led some Swedish consumers to stop buying cars with metallic paint. The many plastic body panels appearing on cars these days often require their own separate paint processes with their own separate problems.

The chemical industry has been trying to solve the problem by making paints more solid and by using water as a thinner. Water-based paints are now being introduced by the motor industry, especially in Europe (GM Opel was the first, at its Bochum plant in 1981), and it is likely that by the middle of the 1990s, most European vehicle plants will use water-based paints almost exclusively, thus removing a major source of pollution. There are still some problems, though. Water-based paint is so different that a totally new paint plant is normally needed. Pipes have to be made of rustproof materials and drying times are longer, which means longer time spent in spray booths and drying ovens. Besides, present water-based paints seem to work best with lighter metallic colours and temperature and humidity need to be more tightly controlled. But water-based paints do still use some VOCs and more paint is wasted in the spraying process.

An alternative solution is under trial at General Motors' Parma, Ohio, plant. This involves painting the steel using rollers before it is pressed. This method does not require the airborne solvents used in spray-painting. The painted steel is baked and coiled. No special stamping techniques are required, but the painted steel cannot be spot-welded in the conventional way. Instead, bonding (i.e. glueing) can be used as can certain mechanical fastening methods (bolts, screws, folding) and trials with laser welding are also taking place. The experiments involve the production of relatively simple exterior panels such as bonnets and bootlids which do not require the complex shapes needed to achieve rigidity in the basic bodyshell, but it seems an important step in the right direction. The prepainted steel technique is already widely used in the production of household appliances like fridges and washing machines.

Several other options are also under consideration. One involves supplying the paint as a thin sheet or film, which is then placed over the welded body and melted in an oven, very much like cheese. Solvent emissions would be minimal and there would not be the waste associated with the spraying of paint. Union Carbide has recently presented a system that replaces two-thirds of the VOCs with carbon dioxide, which would be recycled and reused inside the plant. It is interesting to note that De Lorean solved the whole paint problem by covering the plastic bodies of his DMC 12 sports car with unpainted stainless steel.

Glass, water, fuels, lubricants, etc.

CFCs used in automotive air-conditioning systems are being phased out amid concern over their harmful effect on the ozone layer. Various alternatives are on trial or are being introduced, but none of these has so far proved as effective as the CFCs it replaces. This is already having a major impact on glass technology as glass manufacturers are developing special window glass that filters out a much greater proportion of sunlight than was hitherto possible in order to reduce the temperatures in car interiors.

These special types of glass often require more energy input in their production, while the materials used to block out parts of the solar spectrum may prove more harmful than the usual glass used in cars. The rationale is that if interiors become less hot, a less powerful air-conditioning may be adequate, or perhaps even no air-conditioning at all. Unfortunately, modern aerodynamic car designs often use much larger areas of glass, which allow much more sunlight into the car. This causes various problems including an increase in plastic fumes and more rapid deterioration of plastic parts due to the effect of ultraviolet light (as well as hot drivers and passengers!).

Another important user of CFCs is the electronics industry, which uses these chemicals for cleaning printed circuit boards. This growing industry is an increasingly important supplier to the motor industry and cars could be said to be indirectly responsible for a significant proportion of CFCs released by chemical plants. Few clear alternatives have so far emerged from the electronics sector.

Apart from CFCs, car and commercial vehicle assembly plants also handle very large quantities of potentially harmful fluids such as fuel, engine oil, gearbox fluid and automatic transmission fluid, hydraulic fluids for brakes, clutch, power-assisted steering systems, suspension systems, etc. Leakages do occur and wastage takes place, and such concentrated losses can lead to high levels of pollution in the sewage from a vehicle assembly plant. Once in the water supply, industrial waste can cause a whole series of problems.

However, here too the industry is beginning to clean up its act, especially in countries like West Germany, where there is tough legislation covering industrial waste. Volkswagen is building a sewage treatment facility at its main Wolfsburg plant, which will enable the water to be reused after treatment. The plant uses some 360 million cubic metres of water a year (81 billion gallons) and in the future only around 2% of this will need to be taken in as fresh water. The rest will be recycled. Unfortunately such examples are still fairly rare, especially in such areas as Eastern Europe and the Third World.

Alternative construction methods and materials

There are alternatives to the welded steel construction method. Some of these have been tried in the past, while others continue to be employed today, albeit usually limited to small volume production of specialist vehicles.

Wood

The use of wood is as old as the industry and although it has been replaced by steel and plastic in most cars, wood has always continued to be used for cars and is still in use today, though on a very limited scale. The traditional English quality car with its walnut veneered wooden dashboard is now influencing car interior design worldwide and even the latest Japanese prestige cars like the Toyota Lexus and Mazda/Eunos Cosmo use wood in their interiors. Even as a construction material for cars, wood continues to interest the industry as some recent experiments show. Its main advantage for our purposes is that wood is a renewable resource, provided the woodlands are well managed. It is still unlikely that enough wood could be produced to sustain the present level of world car production and a longer product life as well as much lower production levels would therefore be essential.

Automotive applications of wood have often appeared in response to new developments in wood technology. The Australian Southern Cross car for example, developed by aviator Sir Charles Kingsford Smith and colleague Jim Marks, featured a laminated wood technique as used in aircraft construction, to produce an integral body-chassis structure, streamlined according to the fashion of the 1930s. The car incorporated several advanced features and performed well, partly due to its light body structure. Unfortunately, Kingsford Smith disappeared while attempting to break the UK to Australia air speed record before the project was properly established and although four prototypes were built, the Southern Cross car died.

Several more attempts have taken place and many private individuals have built wooden cars. The original Marcos sports cars of the 1960s, designed by glider and light aircraft designer, Frank Costin, were largely made of plywood and a wooden chassis was retained for production models, but time and cost pressures forced the adoption of a steel chassis later on. Morgans too have traditionally incorporated a wooden floor and various other wooden components in their chassis unit. Wood used in this way has the advantage of being light yet rigid, while it does not rust. Although the wood in Morgans has been known to rot as a result of water ingression, Swedish experience shows that Morgans are surprisingly resilient in a crash.

A UK-designed Third World car, the Africar, was made almost entirely of a resin-impregnated plywood for reasons of low weight, rigidity and the plain fact that plywood is widely available in many areas of the Third

A Nissan Sunny, built out of wood (photo Paul Nieuwenhuis)

World. The production techniques, although relatively labour intensive (not normally a problem in developing countries), require very low levels of investment and this makes it particularly suitable for assembly in isolated areas with simple technology – a key factor of the Africar concept. The UK production plant did not survive, but the concept remains and could easily be revived in a suitable country.

Nissan has also produced a wooden prototype as an employee project (see opposite). This car, essentially a wooden replica of the Sunny, was built partly as a joke but also to see what was possible in constructing a car from wood. As a potential material for car construction, wood is certainly not dead.

Plastics

Plastics are a more popular material than wood and a large number of producers already use plastics instead of steel for their cars. A glassfibre reinforced plastic (GRP) integral body structure was a feature of the Lotus Elite of the early 1960s, but this system is rare, most producers preferring instead to employ a steel chassis to carry the plastic body. These plastics normally involve GRP and increasingly also the very expensive carbonfibre reinforced resins. Most of the small specialist sports car manufacturers use plastic bodies, from Lotus, TVR, Ginetta, Evante and others in the UK, to MVS and Alpine in France and the very expensive and exotic Isdera in Germany. Larger producers use this material for some of their more exclusive low-volume products such as BMW's Z1 sports car, Renault's Espace (developed and built by specialist Matra) and GM's Chevrolet Corvette and now defunct Pontiac Fiero.

Clearly this technique is well established and improvements and further developments are constantly taking place, with even plastic engines being proposed. These may well make plastics more viable for volume-produced cars in the future. Principal suppliers like General Electric Plastics certainly hope so and are investing large sums of money in selling this idea to the car producers. The car producers respond by incorporating more and more plastic parts in their steel-bodied cars.

In terms of production, plastic car bodies require much lower capital investments; no press shop is needed and panels do not need to be welded together by expensive robots. This has always made this technique attractive for low volume production. However, long curing times, large and complex body moulds and expensive resins have limited the appeal to the specialists. This may change as weight reduction and the need for complex shapes in streamlined bodies become more important criteria. Plastic does not rust, but is also difficult to recycle.

Aluminium

Aluminium has a number of disadvantages which make it less attractive than steel for the production of car bodies. It is more difficult to weld, for example, is less rigid for a given sheet thickness and therefore less suitable for integral construction. However, some of the main aluminium producers have recently come up with systems that will allow a car to be built out of aluminium almost as easily and cheaply as from steel. The Honda NSX supercar is largely made of aluminium and Audi have also taken up this idea for the new generation Audi V8.

Until the aluminium industry convinces more volume producers to invest in such systems, aluminium bodywork remains in the realm of the very expensive exotic cars. Bristol and Aston Martin use aluminium bodies, as do some Ferraris, like the F40. These are normally fitted to a steel chassis to achieve the necessary rigidity. A cast aluminium chassis was first shown in 1928 on the French Sensaud de Lavaud prototype. The idea was revived by J.A. Grégoire at the Paris motor show in 1946, but never caught on for production cars, although in the 1950s Saab built six examples of its very light Sonnet Supersports with a chassis structure made up of sections of sheet aluminium riveted together; some racing cars use this technique today. One less expensive car that features aluminium body panels is the Panther Kallista and other small-scale sports cars – like the Caterham Seven – have some aluminium body parts. The Land Rover also traditionally uses an aluminium body.

Aluminium offers similar advantages to plastics. It is lighter than steel and does not rust, although some types of corrosion can occur. One major advantage of aluminium is its recyclability, especially compared to most plastics.

Composite materials

Various combinations (composites) or layers of different materials (so-called sandwich type composites) are now being introduced in car production. These materials often use layered compounds of plastics and aluminium with the resulting material being stronger than either of these materials on their own. The first production car using a body structure of this type (plastic/aluminium/foam) was to have been the Treser TR1 sports car built by Audi-tuner Walter Treser in West Berlin. However, he ran out of money before production began and only a small number of cars were built.

Another car using such a high-tech material was the Panther Solo 2. This was briefly in limited production during 1990 after years of development. The most technically advanced part of the Solo was the central passenger area, which consisted of a steel cage to which a composite (aluminium honeycomb reinforced) sandwich was glued. Kevlar and carbonfibre reinforced plastics were also used in this car to produce one of the stiffest structures on the road. These techniques were pioneered in aircraft production and Formula 1 racing car design. They did form a significant part of the high price of the Solo, as such materials are expensive to produce and the methods for using them to build a car are also expensive and labour intensive.

Clearly then, there are reasons why these various alternatives are not more widely used in volume car production. Apart from being expensive and labour intensive, the manufacturers have invested millions of pounds, dollars, marks or yen in systems that are very good at churning out welded steel bodies by the million, but are not suitable for anything else.

6: *How green is your car?*

Every time a petrol or diesel engine is used, it emits pollutants into the atmosphere. However, the potential level of emissions produced by each model is by no means the same and, within the limitations of the current product range, there is sufficient choice to enable us to minimize the amount of pollution our vehicles cause.

Much attention has been focused on the availability of cars equipped with catalytic converters which, despite their drawbacks, are the method chosen by the UK government to reduce noxious emissions in the short to medium term.

Catalytic converters are a relative newcomer to the UK, and by the end of 1990 only a small number of cars were equipped with them. Nonetheless, 1990 was the year in which sales of cat-equipped cars became significant; in that year 108,422 new cars were equipped with catalytic converters (of which 98,487 were of the three-way type), compared with 9941 in 1989 and a mere 362 in 1988. This still puts the UK far behind such countries as Germany and the Netherlands, where the majority of new cars have catalytic converters, and Sweden, Switzerland and Austria, where cat-equipped cars have been necessary for some years as a result of very strict emissions standards.

After a few early pioneers, we have seen an increasing number of manufacturers offering catalytic converters on their cars, although their policies have differed widely, some showing far more enthusiasm for catalytic converter technology than others. By 1991, nearly all manufacturers present in the UK market were making catalytic converters available, either as standard equipment, as a cost option, or as a no-cost option, on at least a selection of their models.

Not surprisingly, sales of cat-equipped cars in the UK were spearheaded by West German, Swedish and Dutch manufacturers, notably Volkswagen-Audi, Volvo and Saab. Although it was Toyota who first offered a car with catalytic converter as standard in the UK, Volkswagen claims to be the first manufacturer to have offered catalytic converters as an option on cars sold in the UK, and by the end of the 1980s versions of almost all Volkswagen models could be purchased with catalytic converters as standard equipment. However, Volkswagen's partner Audi went one better and has been supplying all its cars with catalytic converters fitted as standard since the beginning of the 1990 model year. Volvo chose a slightly different approach at first, offering catalytic converters as a no-cost option on some models and as standard equipment on others, before adopting standard availability across the board for the 1991 model year.

By the beginning of 1991, the biggest manufacturers in the UK market had not opted for catalytic converters as standard equipment on all their cars, partly because of the effect this might have on the price of their cars. Indeed, with some exceptions, it was the low volume, upmarket manufacturers which had embraced the catalytic converter most readily, as they were more willing to absorb the costs of incorporating the equipment. Despite the big gaps that still existed in 1991, all new cars have to be equipped with catalytic converters from the beginning of 1993, with many manufacturers providing them as standard equipment before this date and some manufacturers waiting for the new emissions limits to come into force.

Milestone of cat availability

The following list shows which manufacturers achieved the first breakthroughs in cat availability (some of them more symbolic than comprehensive) in the UK between 1988 and the end of 1990.

Toyota

Toyota was the first manufacturer to equip one of its UK-spec cars with a catalytic converter as standard (Celica GT4, March 1988).

Volkswagen

Volkswagen was the first volume manufacturer to offer a catalytic converter as standard equipment on a small car (autumn 1989). This was on the Polo (although only one version of the model was thus equipped). Volkswagen was also the first manufacturer to offer a catalyst retro-fit policy in the UK.

Vauxhall

Vauxhall was the first volume manufacturer to undertake to equip all its new cars with catalytic converters by a certain date. It announced in April 1989 that it would fit all its new cars with regulated three-way catalytic converters by spring 1992.

Saab

Saab was the first manufacturer to equip cars for sale in the UK with a climate-adjusted regulated three-way catalytic converter. The climate-adjusted system, available on all Saab's 16-valve models since September 1990, was designed specifically to operate in north European conditions, in that it would work at 2 degrees centigrade (most catalytic converters have been designed for optimum efficiency in Californian conditions, where start-up temperatures are situated between 20 and 30 degrees centigrade).

Audi and Porsche
Audi and Porsche were the first manufacturers to equip all their cars sold in the UK with catalytic converters as standard equipment (autumn 1989).

The following manufacturers had equipped all their petrol-engined cars sold in the UK with catalytic converters as standard by the end of 1990:
Audi (autumn 1989); BMW (autumn 1990 – except the 3-series model, replaced in the first half of 1991); Jaguar (autumn 1990 – except the XJS 3.6); Porsche (autumn 1989); Saab (autumn 1990); Volvo (autumn 1990).

Beyond the catalytic converter

A tried and trusted technology, catalytic converters are at present the most effective means of reducing noxious emissions from conventional engines, and alone can meet the EC emissions standards of the early 1990s. However, they do not help reduce output of carbon dioxide, the principal greenhouse gas. On the contrary, they may even exacerbate this problem inasmuch as they require a relatively rich fuel-air mix in order to operate properly and they speed up the conversion of carbon monoxide into carbon dioxide. Lean-burn engines (see Chapter 2), which work with a smaller ratio of fuel to air, have already been developed, but as yet their emissions (particularly of nitrogen oxides) are still too high to enable them to conform to the strictest environmental standards unless a catalytic converter is fitted.

Lean-burn engines already installed in production cars include Ford's HCS (High Compression Swirl) engines, which power the Fiesta (1.1 litres) and Escort (1.3 litres). Peugeot has developed a larger lean-burn engine (1.9 litres) called CERES or, more technically, XU9 J2 Lean Burn. This was tested on the Peugeot 309, Peugeot 405 and Citroën BX, and surpassed the EC emissions standards which were originally to have been introduced in 1991-92, but which have since been rejected in favour of stricter controls. Toyota also offers a lean-burn engine on its Camry models in some markets.

Honda and Mitsubishi both revealed new fuel-efficient engines in the latter half of 1991. The new engines, which are both of 1.5 litres capacity, do not represent a technological breakthrough, but are based on the improvement of existing techniques. The Honda engine combines a high-swirl combustion chamber with a lean-burn fuel system and a computer-controlled variable-valve timing mechanism (VTEC-E), which allows a more precise air-fuel mix in the combustion chambers. Honda claims to have achieved an air-fuel ratio of 25:1 with its new powerplant, leaner than the mix achieved by Peugeot with the CERES engine. Installed in the three-door Civic VEi, and fitted with a three-way catalytic converter, the new engine is said to return 62.8 mpg at 56 mph, which should make it the most fuel-efficient 1.5 litre petrol engine available in the European market.

Impressive though this performance may be, the problem for the new Honda and Mitsubishi engines, as for all lean-burn units, is that they are prevented from achieving their full potential by the need to fit catalytic converters, which will not function effectively if there is too much oxygen in the exhaust.

Most car manufacturers consider that, in the absence of radical emissions legislation outside California, petrol and diesel engines will continue to be the principal sources of motor vehicle propulsion. Research is being carried out by manufacturers on more effective lean-burn engines, which it is hoped will eventually be able to meet US-equivalent emissions standards without the need for catalytic converters (it is impossible to predict accurately if or when this goal will be achieved). One avenue of research is the direct injection petrol engine, as installed by Volkswagen in its experimental car, the Futura.

Direct injection engines (which are already used in diesels) ensure a more precise use of fuel than indirect injection engines. The level of pressure on the accelerator pedal determines the amount of fuel which is injected into the combustion chambers, whereas in indirect injection engines the accelerator only alters the volume of air, the fuel system adding petrol to maintain an accurate fuel/air mix. Direct injection petrol engines, like their diesel counterparts, operate with high compression ratios, and one problem has been to develop a pump which can deliver petrol at the necessary pressure. Another problem has been to develop spark plugs which can tolerate high temperatures and pressures in the combustion chamber.

Fuel efficiency

A car's 'greenness' depends not only on the amount of toxic gases and particles it emits, but also on the amount of carbon dioxide it releases into the atmosphere. The less fuel a car uses, the less carbon dioxide it produces.

Around 60% of the new cars sold in the UK are, in broad terms, medium sized, i.e. the size of a Ford Escort/Vauxhall Astra/Rover 200 Series (lower medium) or Ford Sierra/Vauxhall Cavalier/Rover Montego (upper medium). A significant minority of new registrations – over a quarter – are of small cars (e.g. Ford Fiesta/Vauxhall Nova/Rover Metro), and the remainder are large cars.

On the whole it is true to say that small cars are more fuel-efficient than larger ones, but this is not a rigid rule and there are a number of exceptions, many of which result from the relatively recent trend of installing larger and more powerful engines in small cars. Small engines also tend to be more sparing on fuel than large ones, but there are some exceptions here as well, for installing a small engine in a large car may in some cases mean that the engine has to do more work, thereby increasing emissions. Technological progress means that an existing engine may be outperformed by a new, larger and more fuel-efficient engine introduced by the same manufacturer.

It is indisputable that diesel engines are currently more fuel-efficient than their petrol counterparts; as will be seen below, the least thirsty cars in the UK market are small diesel-powered cars, and there are a number of fairly large diesel cars which easily outperform smaller petrol-driven vehicles in this domain. Diesel cars take a relatively small share of the UK market, despite the fact that diesel offers lower fuel consumption. This situation has occurred as a result of a combination of interrelated factors – relatively low petrol prices, the existence of only a small price differential in favour of diesel, and the traditional perception of diesel as dirtier and more noxious than petrol. This view is being challenged by manufacturers which produce a large number of diesel cars – they argue that diesels emit as few, or fewer, hydrocarbons and as little, or less, carbon monoxide than cars fitted with a catalytic converter (see Chapter 2).

In some European countries, such as France and Italy, diesel fuel has enjoyed a wider price differential in its favour than in the UK. This has helped diesel cars to achieve a substantial share of the market (20-30%) in those countries, whereas in the UK diesels accounted for just over 6% of new car sales in 1990. Nonetheless, this represented a record level for diesel sales in the UK (sales went up from 123,000 in 1989 to 128,000 in 1990, despite a fall in total new car sales of more than 12%). Demand for diesel cars grew most steeply in the last quarter of 1990, when a series of petrol price rises took place in response to events in the Gulf. When fuel prices rise, the inherent fuel economy benefits of diesel cars become a higher priority.

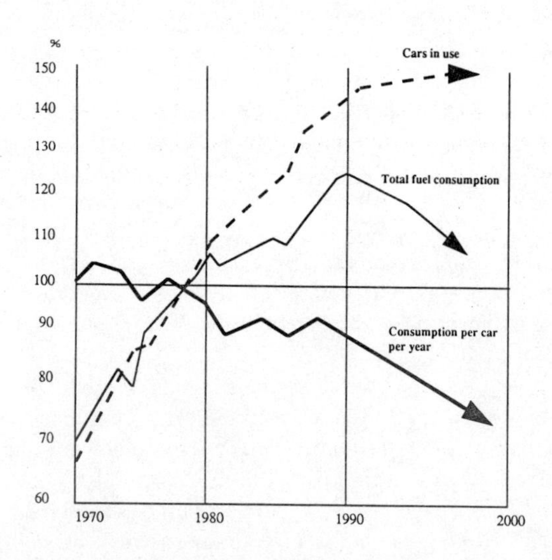

Annual fuel consumption per car in West Germany (1970 = 100%) compared with the number of cars in use and total fuel consumption (*Source: VDA*).

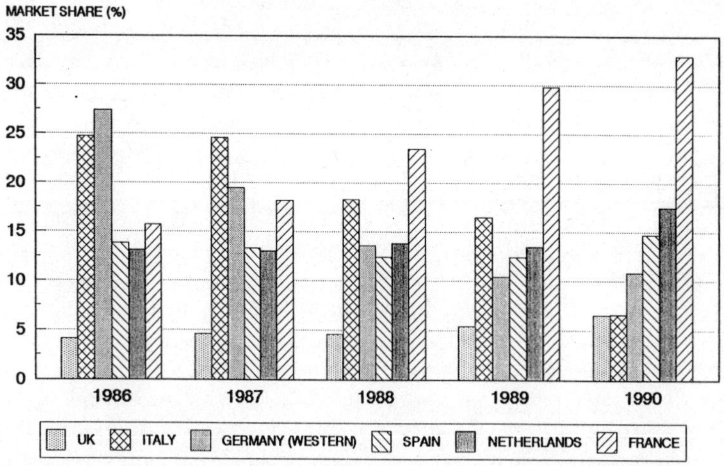

Diesel car sales in major European markets 1986-90 (% of total car sales)

Most cars sold in the UK have an mpg rating in the thirties or forties. This statement provides no indication of fuel efficiency trends, which is best illustrated by the graph opposite showing fuel efficiency trends in the West German market.

The tables used in this chapter to rank cars according to their fuel efficiency refer to the situation in October 1991 and use the results of official government fuel consumption tests. Three measurements are employed – fuel consumption on the urban cycle (i.e. typical town driving conditions), fuel consumption at 56 mph, and fuel consumption at 75 mph. This gives a more detailed picture than a simple average rating.

The most fuel-efficient cars sold in the UK

Model	Fuel consumption (mpg)		
	urban cycle	*56 mph*	*75 mph*
Citroën AX14 TGD/DTR	54.3	78.5	56.5
Daihatsu Charade CX DT	57.6	78.5	49.6
Peugeot 205 1.8 D/XLD/GLD/GRD	52.3	72.4	54.3
Citroën AX10 TGE	50.4	72.4	50.4
Rover Montego 2.0 DSLX Turbo	48.6	75.0	51.1
Citroën AX11 TGE	49.6	72.4	50.4
Volkswagen Passat CL TD	42.2	72.4	56.5
Ford Fiesta 1.8D/LXD	48.7	70.6	50.4
Vauxhall Nova 1.5 TD	48.7	70.6	50.4
Renault 5 Campus 5-Sp Cat	48.7	68.9	50.4

Of the above ten models, eight are diesel-powered cars (identified by the letter D in the model designation) and eight are classed as small cars. Surprisingly, perhaps, two upper-medium segment cars – the Rover Montego and the Volkswagen Passat – score highly in the fuel efficiency league table, which demonstrates that exceptionally fuel-efficient cars do not have to be small, so long as they are powered by diesel engines. The Montego is powered by a direct injection diesel engine.

The Citroën AX diesel, powered by a 1361 cc engine, just beats the turbo-assisted Daihatsu Charade to become the most fuel-efficient car available, but two of its petrol-engined sisters, the AX10 and AX11, are also very well placed. Citroën achieves its record partly through the lightness of the AX's construction – at 640 kg it is lighter than all other cars in the same class. The AX also has a very low drag coefficient (Cd 0.31), which makes it the most aerodynamic of all production cars below 3.7 metres in length.

Among medium-sized cars, diesels are clearly the most fuel-efficient. The best performers, led by the Montego 2.0 turbodiesel, appear in the table above, but the following list provides a fuller picture of the most fuel-efficient medium-sized cars.

The most fuel-efficient medium-sized cars sold in the UK

Model	**Fuel consumption (mpg)**		
	urban cycle	*56 mph*	*75 mph*
Rover Montego 2.0 DSLX Turbo	48.6	75.0	51.1
Volkswagen Passat CL TD	42.2	72.4	56.5
Vauxhall Astra 1.7 D Merit 3-dr	44.1	70.6	49.6
Rover Maestro 2.0 DLX	47.7	64.5	45.8
Ford Escort 1.8 D/LD	47.1	67.3	50.4
Citroën BX 17 TGD/TZD	44.1	61.4	44.1
Audi 80 Turbo Diesel	40.4	65.7	47.9
Citroën BX 19 TGD	43.5	61.4	46.3
Rover 218 SD	42.2	63.2	47.9
Vauxhall Cavalier 1.7 DL/GLD	42.2	64.2	47.1

All of these cars are diesels, they are more fuel-efficient than a number of petrol-engined small cars, and they easily beat medium-sized cars powered by petrol engines. It is also notable that a relatively heavy car, the Volkswagen Passat, scores very highly in the above table. This serves to demonstrate that individual criteria should not be accorded too much importance in assessing the relative fuel efficiency of different models – this, on the contrary, results from the overall design of a vehicle, and is determined by the interaction of all powertrain and structural elements.

Unlike on the continent of Europe, many manufacturers – such as BMW or Alfa Romeo – do not offer diesel versions of their larger cars in the UK. Those manufacturers that do, however, can offer fuel economy from large cars which matches that of some small petrol-engined vehicles, reinforcing the fact that diesel engines provide unrivalled fuel efficiency.

On the whole, it is the Citroën XM turbodiesel which emerges as the most fuel-efficient large car. This is achieved through the use of multivalve technology – the XM turbodiesel is powered by a four cylinder, twelve valve engine (two valves per cylinder for the inlet and one for the exhaust). Multivalve technology enables the XM to be both more fuel-efficient and more powerful than the turbodiesel Citroën CX which it has replaced, and its fuel economy is also promoted by a lighter structure and better aerodynamics.

The Audi 100 turbodiesel, like the Rover Montego, is powered by a direct injection engine, demonstrating the high level of fuel efficiency that can be achieved with such engines in relatively large cars.

The most fuel-efficient large cars sold in the UK

Model	**Fuel consumption (mpg)**		
	urban cycle	*56 mph*	*75 mph*
Citroën XM Turbo SED	33.2	57.6	43.5
Peugeot 605 SR dt	33.2	57.6	43.5
Audi 100 Turbo Diesel	33.6	55.4	39.2
Vauxhall Carlton 1.8 L	28.8	50.4	39.8
Volvo 940 TD	31.0	48.7	35.3
Ford Granada 2.0i LX/GL/Ghia/Ghia X*	28.8	45.6	39.2
Renault 25 GTS	26.9	49.6	37.2
Rover 825 D Turbo	33.8	57.8	45.2
Vauxhall Carlton 2.3 LD**	35.8	54.3	40.4
Fiat Croma 2.0 CHT	30.7	51.4	39.2

*Ford Granada Ghia Cat and Ghia X Cat have figures of 29.1/46.3/37.2
**Estate version has figures of 35.8/52.3/37.7

At the most rarefied end of the market, represented by the fastest sports cars, or supercars as they are sometimes called, fuel efficiency is not a major demand at present. However, some manufacturers recognize that future supercars will be significantly more fuel-efficient, possibly because legislation will demand it. One of these manufacturers is Lotus, all of whose cars are currently powered by four cylinder engines. An absence of official fuel consumption figures for a number of sports car manufacturers makes it impossible to compare the fuel efficiency of the various companies' products, but one factor in Lotus's favour appears to be the relative lightness of its models, achieved through the use of synthetic material for the body and a light engine (only 4 cylinders).

Recycling used cars

The question of recycling of materials and components from old or damaged cars has recently gained a high profile. As pointed out in Chapter 4, a car's metal content (steel, cast iron, zinc, copper, etc.), which represents

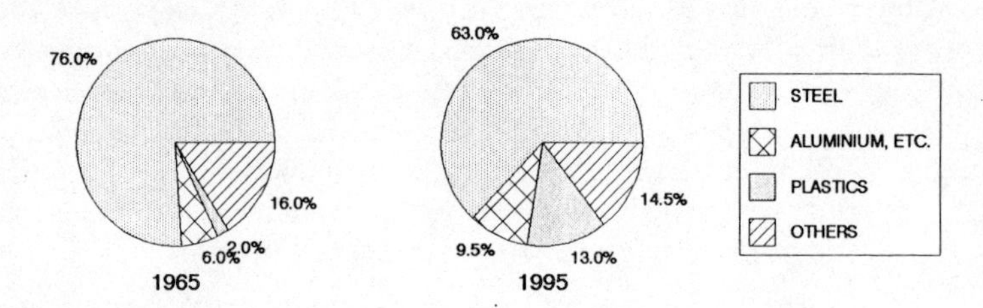

Materials used in a car 1965 (actual) and 1995 (forecast)
(Source: Mercedes-Benz)

on average 75% of its weight, is recycled on a large scale. Other materials – plastics, fabric, rubber, glass, etc. – have posed a greater problem, and the prevailing practice is still to put them through a shredder and then dump them at waste sites.

The problem is assuming greater importance, as plastics in particular are finding wider applications in motor vehicles. Led by the Germans, many car manufacturers are now taking a serious interest in recycling these more difficult materials. Some plastic components are already being recycled – Mercedes-Benz, for example, has a pilot project for the recycling of bumpers, and the floor mats in the latest BMW 3-Series car use material made from recycled 5- and 7-Series bumpers. Furthermore, plastic body parts from the low volume BMW Z1 are returned to the supplier in the event of damage and granulated, producing the basic material for new components.

Recycling has become a significant factor in the marketing of new cars. On the launch of the 3-Series saloon in 1991, BMW's publicity arm emphasized the recyclability of its plastic components, and went so far as to distinguish those parts which were capable of being recycled from those which were already being made of recycled material. The former were clearly in the majority, but one of the aims of BMW's pilot recycling programme in Germany is to enable larger scale recycling to be carried out on an economic basis.

All the major car manufacturers in Germany are pursuing plans to recycle old cars on a comprehensive, large scale basis. Volkswagen has its own pilot recycling project at Leer, close to the Emden car assembly plant, and Opel has set up an association grouping together companies in the Ruhr region of Germany which aims to achieve nothing less than 100% recycling of cars. The large French manufacturers, Renault and Peugeot, are following the example by inaugurating their own pilot recycling programmes, although they also intend to continue collaborating with each other on this subject (see Chapter 4).

Recycling is also coming to the UK, with BMW again in the forefront. A pilot programme is to be established by the end of 1992, in co-operation with the Motor Vehicle Dismantlers' Association, and the lessons learned will be incorporated into a national network of car dismantling and reprocessing centres. Five such centres are planned for the end of 1993, increasing to ten by the end of the following year. The aims are to achieve the maximum recycling potential of BMW cars on an economic basis and to ensure that all such cars, on being discarded by their owners, are recuperated and recycled in the UK.

In October 1991, BMW began collecting and shredding damaged bumpers from the 3-Series car in the UK. The resulting material is shipped to Germany and reused in floor mats, boot linings and other components. In a similar way, damaged catalytic converters are collected and sent to Germany, where the precious metals that they contain – platinum and rhodium – are recovered. The car owner is offered a rebate towards the fitment of a new catalytic converter.

BMW dealers in the UK are being made responsible for the recovery of certain components and waste products, such as batteries, oil, brake fluid, paint waste, glass, solid consumables, parts cleaning equipment and paper, cardboard and boxes. A series of schemes were established by the end of 1991, involving specialist recycling companies.

While many manufacturers are developing ways of making their cars more easily recyclable and setting up the procedures that will make large-scale recycling possible, Volkswagen has stolen the limelight by undertaking to recycle its latest Golf model. The cars will be recovered from the owner free of charge and sent for recycling, although it is not yet certain just how much of the material will be reused. The programme will commence in Germany, but is expected to cover all European countries eventually. Following Volkswagen's announcement, Opel has expressed an intention to take back the Astra model for recycling. Given that both the Golf Mk III and the Astra are new models, the volume of cars for recycling will not rise steeply until after the mid-1990s – by then, the systems for large-scale recycling should be in place.

'Green' features by manufacturer

The following list attempts to cover some of the environmental record of the main producers selling in the UK. This list cannot claim to be complete as many research projects go on all the time.

Audi

In autumn 1989, Audi was one of the first manufacturers to supply catalytic converters as standard on all its new cars sold in the UK. The company has stated that the Audi 80 turbodiesel, powered by a 1.6 litre engine, and the Audi 100 turbodiesel, powered by a 2.5 litre direct injection engine, conform to US 1983 standards. In terms of alternatives, Audi showed a simple hybrid petrol-electric car at the 1990 Geneva motor show.

Audi is taking a serious interest in weight reduction. Its recent 'supercar' concept, the Spyder, was designed to demonstrate the potential of aluminium for weight saving in car construction. The car has a body and frame made entirely of aluminium, which means that, despite being powered by a 2.8 litre V6 engine and equipped with four-wheel drive, it is lighter than the entry-level Audi 80 saloon. For large saloon cars, Audi is said to be aiming at a weight reduction of 20% compared with current levels. To this end, the next V8 saloon, scheduled for launch in 1993, will make significant use of aluminium.

BMW

BMW is experimenting with a hydrogen engine and is involved in the AGATA gas turbine project, entailing the collaboration of a number of European car manufacturers under the aegis of the EC's Eureka programme. Daimler-Benz, Volvo, Volkswagen, Peugeot, Renault and Fiat, together with Garrett (a turbine producer) are also involved. BMW has also been experimenting with electric versions of its small 3-Series saloon. It is committed to recycling and has established a purpose-built car reprocessing plant at Landshut in Bavaria.

BMW has produced an electric-powered concept car, the E1, which reveals the company's thinking on the appearance of future electric cars and on the technical solutions they will contain. The rear-drive E1 is only 3.4 metres long, but offers excellent interior space utilization. It contains high-energy sodium-sulphur batteries which power a 32 kW electric motor, and has a top speed of 120 kmh. The E1 can travel 200 km between battery charges, but in city driving this would be reduced to 150-200 km, which nonetheless represents an improvement over its earlier experiment with a battery-powered 3-Series saloon. The bodyshell is made of aluminium with an outer skin of recyclable plastic.

Citroën

If one manufacturer has to be singled out as having the best record on fuel efficiency, it probably has to be Citroën. Not only does Citroën offer the most fuel-efficient production car available in the UK market, but it has the most economical petrol-engined car, according to available figures, and its medium-sized and large models are also well represented in the fuel economy stakes. Citroën offers a number of electric versions of its vans. (See also Peugeot).

Citroën was carrying out studies on gas turbine engines as early as 1958, and research was pursued later in collaboration with Fiat.

Fuel-efficiency tests were carried out in the 1980s under the aegis of the ECO 2000 programme, with the aim of covering a distance of 100 km using only 3 litres of fuel. The results of this programme were fed into the development of the Citroën AX, which is the most fuel-efficient car in large-scale production.

Daihatsu

Daihatsu is actively involved in electric vehicle development.

Fiat

Fiat launched an electrically powered version of the Panda (called the Panda Elettra) in Italy in summer 1990. The Elettra is most suited to urban use – it can travel over 50 miles on fully-charged batteries and has an initial top speed of 45 mph, which can be increased to 75 mph. Fiat claims that this is 'the first electrically-driven car produced in volume by a major manufacturer'. The Italian company is also a signatory to the AGATA gas turbine project.

Ford

Ford of Europe has invested considerable sums in lean-burn research. Its US parent joined with GM and Chrysler in early 1991 to work on the development of battery technology for future electric cars.

Ford has been collaborating closely with the Orbital Engine Company on the development of a two-stroke engine which it may eventually install in a new car smaller than the current Fiesta.

Honda

Honda is among the pioneers of emissions technology, dating right back to the then clean CVCC engine of the early 1970s. Its engine expertise has led to greater efficiency from smaller engines using multi-valve heads and variable valve timing (V-TEC). The new Honda NS-X sports car uses an all-aluminium body for lightness.

Honda has installed a new lean-burn engine in its latest Civic model which may make it the most fuel-efficient petrol-engined car in its class. The engine is a 1.5 litre SOHC unit, the VTEC-E, and is said to achieve 62.8 mpg, or 42.2 mpg on the urban cycle.

Jaguar

Natural, renewable materials such as leather and wood are a feature of Jaguar interiors.

Land Rover

Land Rover has been using aluminium for a long time in some of its vehicles, and its products are noted for their longevity.

Lotus

Lotus cars are particularly light as a result of their construction and GRP bodies; this also makes them resistant to rust and helps in extending their useful life. Representatives of the company have mentioned the importance of producing lighter and more fuel-efficient cars in the future. Lotus is even working on a solution to the noisiness associated with light cars; it has been developing a system which generates a sound signal to cancel out noise from various sources in the car, including engine and tyres. The system may eventually make it possible for luxury cars to dispense with noise-reduction features such as large engines and heavy body panels, which militate against fuel economy. It could also give small cars the same refinement as large cars, thereby removing one of the advantages of larger cars over smaller ones.

Mazda

Like other Japanese producers, Mazda is actively involved in the development of electric vehicles.

Mercedes-Benz

The quality and longevity of Mercedes cars is legendary, and in Germany (though not in the UK) the company was among the first to offer a catalyst as standard equipment. Mercedes-Benz is involved in the AGATA gas turbine project and has been experimenting for some time with hydrogen power in cars, vans and buses. Mercedes has also invested in improved diesel engines with positive results in terms of emissions and noise, and has launched a major recycling programme.

Mitsubishi

Mitsubishi has carried out work on electric vehicle technology, and has established a joint venture with the Tokyo Electric Power Company to develop an electric vehicle using nickel-cadmium batteries.

Nissan

Nissan is involved in gas turbine and electric vehicle research. Nissan is working with Kyushu Electric Power on an electric car which it wants to be able to cover the same distance as a petrol-engined model. The requirement is to achieve over 200 km at a speed of 40 kmh, using only one battery charge. A prototype of such a car, called the Future Electric Vehicle (FEV), has already been presented. This uses nickel-cadmium batteries which are significantly smaller and lighter than conventional cells, and it also has solar panels on the roof. Recharging the batteries takes just 15 minutes, a huge improvement over traditional battery technology.

Peugeot

Peugeot S.A. (which includes Citroën) has spent considerable resources on lean-burn research. The group is producing an electrically powered version

An electric-powered Peugeot 205 (photo PSA)

of the 205 model at its Mulhouse plant for sale in France, and has also made available electric versions of its C15, C25 and J5 vans. Peugeot is also a member of the AGATA gas turbine project and is experimenting with a number of other alternative power sources, most recently in a joint programme with Renault, funded partly by the French government. Among other things, the programme involves research into two-stroke engines, electric cars, hydrogen technology and vehicles powered by natural gas.

Peugeot has set up an experimental joint venture with a scrap metal processing firm and a cement maker to recycle old vehicles. The operation entails the recovery of materials such as plastics, rubber and glass which are usually disposed of in landfill sites in the form of 'shredder fluff'. The venture began in July 1991 at a plant in Saint-Pierre de Chandlieu, France, and it is expected that around 7200 Peugeot, Citroën and Talbot vehicles will go through the recycling process over a two-year period. The pilot project is intended to help Peugeot design vehicles which are easier to disassemble and to determine which recycling methods are economically feasible on a large scale.

Porsche

Porsche was one of the first manufacturers to supply catalytic converters as standard on all its UK-spec cars (autumn 1989). The company carries out a large number of research projects, such as the long-life car programme and the use of metal catalytic converters, which reach their operating temperature faster than ceramic converters. All Porsche cars feature fully galvanized steel bodies for durability. High performance four-cylinder engines are used in all 944 models (see also Lotus).

The Renault Vesta II (photo Renault)

Renault

In 1987, Renault achieved the world record for low fuel consumption in normal traffic conditions (1.94 litres per 100 km, at an average speed of over 100 kmh) with a car produced under the Vesta project. The lessons learned from this project have been used in Renault's product development policy, notably in the Clio, launched in the UK in 1991.

Renault is involved in the AGATA gas turbine project and is working on a number of other alternative fuel programmes in conjunction with Peugeot, with whom it is also collaborating on the recycling of materials from old cars.

Rolls Royce

Rolls Royce and Bentley cars have a particularly long average life expectancy and the firm supplies parts for most older models to help keep them roadworthy.

Rover

Rover deserves a special mention for achieving the highest level of fuel-efficiency for medium-sized cars, exceeding that of many smaller diesel-powered vehicles.

Saab

Saab pioneered work on the turbocharging of four-cylinder engines, with the aim of achieving higher power output while avoiding a deterioration in fuel economy. It is the only company in the executive segment which relies exclusively on fuel-efficient, low-emission four-cylinder engines, where most competitors have six cylinders.

The Toyota AXV-IV, a lightweight 'commutercar' powered by an 804cc two-stroke engine (photo Peter Cope)

Toyota

Toyota is pursuing research into lean-burn technology and gas turbines. The gas turbine engine, installed in an experimental vehicle called the Toyota GTV, can meet all current emissions control requirements without recourse to a catalytic converter, and can use different fuels, including methanol and ethanol. Toyota also has a comprehensive programme for the development of electric vehicles.

Toyota has been working on both petrol and diesel two-stroke engines, and may be one of the first manufacturers to install this technology in its production cars.

Toyota has produced a prototype of a commuter car made from lightweight materials (aluminium, magnesium and plastics reinforced with carbon-fibre). Dubbed the AXV-IV, the car is 340 cm long, weighs only 450 kg, and is powered by a 64 bhp, 804cc two-cylinder two-stroke engine employing five valves per cylinder. The engine itself weighs a mere 83 kg.

Toyota favours methanol as a future alternative to petrol engines, and is working on flexible fuel (methanol-diesel) technology.

Vauxhall

Vauxhall's US parent has tested and exhibited a purpose-designed electric car known as the Impact (see page 93). This can travel at 100 mph and has a range of 120 miles on one battery charge. It also has a compact current converter to change DC to AC current. General Motors aims to carry out volume production of the Impact (to this end it has joined forces with Ford and Chrysler in order to develop new battery technologies), and is looking into the question of tax relief for the car in order to make it more cost-competitive.

General Motors is one of the companies involved in the development of the Orbital two-stroke engine, which is smaller and lighter than a conventional four-stroke unit.

Volkswagen

Aside from the early provision of catalytic converters on petrol-engined cars, Volkswagen has been active in the development of clean diesels. For the 1990 model year it introduced its 'Umwelt' or 'environmental' diesel engine in the Golf and Jetta, as a direct replacement for its old diesels.

Volkswagen has developed a diesel-electric hybrid engine. This is powered alternately by the diesel engine and the electric motor, the latter largely for use in town. A Hybrid Golf will be sold commercially during the first half of the 1990s. Volkswagen has also produced a small purpose-built hybrid car, the Chico, powered by a 636cc two-cylinder engine and a 6 kW electric motor. The car is said to be capable of 81 mph and to achieve 80 mpg in petrol mode and 88 mpg when both engines are in use. It has been designed as a prototype of a car that is to be built in collaboration with the Swiss watch manufacturer, Swatch, which has commissioned VW to develop an electric car with a performance similar to that of a traditional combustion engined vehicle.

Volkswagen is working on direct injection petrol engines and is a member of the AGATA gas turbine project.

The company has become the first European car manufacturer to appoint a board member to deal specifically with environmental matters.

Volvo

Volvo cars are noted particularly for their build quality and longevity, and the firm has a declared environmental policy with, among other features, a commitment to recycling. The company is one of the participants in the AGATA gas turbine project.

7: *The alternatives*

The virtual monopoly of the internal combustion engine using the four-stroke cycle is in fact a relatively recent phenomenon. Cars with the alternative two-stroke cycle (Wartburg, Trabant) were still being made in 1990 in the former German Democratic Republic, although even there pollution and fuel consumption problems caused their final demise.

If we go back to the first half-century of the modern motor car, we find a number of technologies side by side and offered to the consumer as true alternatives. Up until the 1930s, the car buyer with enough money to spend could choose not to go for a petrol engine but to opt instead for steam power or electricity. The first car to go a mile a minute was electrically propelled (Jenatzy in 1899), while the first car to go two miles per minute was steam powered (Stanley in 1906). For various reasons – which are not necessarily as valid now as they were in the past – the development of the motor car is characterized by a gradual decline in the amount of choice available to the consumer.

If we look back further, we find that the earliest vehicles were powered by animals, the horse-drawn vehicles we are still familiar with, but experiments also took place with other sources of motive power, notably wind, although human power and clockwork systems were also tried.

Human power

The Romans used slaves to power large ceremonial vehicles, but in later centuries more civilized technologies prevailed. Human-powered vehicles (HPVs) are known to have been developed by several people from the Renaissance onwards. Giovanni de Fontana (1395-1455) developed one in Padua, Italy. This single person vehicle was propelled via ropes wound around pullies. Steering was not provided. A print by Albrecht Dürer of 1526 shows a human powered vehicle built for the emperor Maximilian I, although its designer is not known. Johann Hautsch is known to have built two different human-powered vehicles (1640 and 1649) in Nürnberg, which were used on official occasions. The later one was subsequently bought by the King of Sweden, whereupon the King of Denmark immediately ordered a similar vehicle! Also in Nürnberg, Stephan Farffler (1670) built two very light human powered vehicles, to overcome his disability – a fall in childhood having paralysed his legs – and he is known to have used them regularly. Later on, the Russian Ivan Kutbin (1752) also built a three-wheeled HPV, which still survives.

In the next two centuries, human power really took off in the form of the bicycle, which is still an ideal vehicle for carrying one person and a small amount of cargo over short to medium distances within, for example, an urban or suburban environment. The bicycle has become deservedly popular and this popularity is still growing. But the bicycle does have disadvantages. When it is combined with faster and heavier forms of traffic, the cyclist is at a distinct advantage and can be placed in a very dangerous position (see Chapter 4). For the bicycle to operate as a serious and safe alternative form of transport, it needs to be separated from other forms of traffic. Another disadvantage is the lack of weather protection. Various solutions to this problem have been developed, mainly in the form of alternative HPVs.

The development of the three- or four-wheel human-powered vehicle did not stop with the rise of the bicycle. There have even been some car manufacturers who have experimented with these. Gabriel Voisin, for example, the great French aircraft pioneer and probably one of the most original car designers ever, built a small HPV in the early 1940s. This 'Vélogab' was primarily for his own use and was propelled by a system that resembled a cross between an old-fashioned sewing machine and a pedal car. Voisin did apparently consider its viability for series production, but in fact fitted it with an electric motor and turned his attention to minimal motorized personal transport.

Ballantine (1988) points out that 'a cyclist moving at 20 mph displaces some 1000 pounds of air a minute, a task that consumes about 85% of the rider's total energy output.' In aerodynamic terms, a recumbent cycle has about 20% less drag than an upright bike, because of its reduced frontal area and overall shape. These characteristics are further enhanced by fitting a streamlined bodyshell, which can reduce drag by up to 80%! This amounts to some 70% less energy consumption for the same speed as well as a much higher overall top speed. The current speed record for an HPV stands at more than 65 mph or over 100 kph!

After early successes in the 1930s, a racing ban caused a halt to HPV development and the conventional bike enjoyed further refinement. It was not until 1975 that the cause of the HPV was taken up again by the formation of the International Human-Powered Vehicle Association (IHPVA) in California. Under the aegis of the IHPVA, the land-based HPV has seen a revival. Human powered hydrofoils have also been built and a human powered aircraft, the Gossamer Albatross, has crossed the English Channel. The wheeled HPVs inspired by the organization have been used mainly for racing and record-breaking.

With these sorts of performance levels, and the weather protection that a faired-in body provides, HPVs are beginning to look like serious alternatives to the motor car, especially for short to medium distance commuter use (up to 20 miles or so). The vehicles are non-polluting, use a minimum of energy (food) and can be made to last a very long time. Considering that

Bicycle: drag consumes 85% of rider's energy output

Recumbent cycle: 20% less drag

H.P.V.: 80% less drag

Energy advantage of the human powered vehicle (H.P.V.)

most car commuting still involves one person travelling in a vehicle designed to take between two and five, or even more, the single seater HPV should present no problems for this type of use. Their limited size also leads to less congestion on the roads and a lower demand for car parking facilities.

Some road-going HPVs have also been developed, among them the Windcheetah SL (= street legal), built in Norwich. This vehicle is available with a streamlined body as an HPV, or without it as a recumbent bicycle (see illustration). These vehicles have proved very competitive in racing and yet are quite suitable for ordinary road use. The most recent competitor in the HPV market is the Dutch Solo. This vehicle, created by Witkar designer Luud Schimmelpenninck (see Chapter 8), was conceived specifically as a commuter vehicle, and by early 1990 several employees of the Dutch traffic department were using them as a practical experiment. Dutch conditions, with a widespread network of cycleways, are particularly suitable for safe and fast HPV commuting. The Solo is more upright than the Windcheetah and this may make it more acceptable to car users.

Some of the advantages (and disadvantages) of the bicycle also apply to motorcycles, although they still pollute – especially those fitted with two-stroke engines – and use up scarce resources. Human power, however, is only one step removed from animal power. We now turn our attention to the first relevant example of people harnessing the powers of nature for propulsion – the wind-powered landyacht.

Wind power

Wind power is of particular interest as it is a 'free' and totally renewable resource. The earliest record we have for wind-powered vehicles is from Egypt around 1830 BC, where the pharaoh Amenemhet III allegedly used one for travelling along the Nile. The Chinese are also known to have experimented with such vehicles and it is this information which inspired the first European experiment.

Dutch seafarers brought back stories of wind-powered vehicles in the sixteenth century and one of these, the travelogue by Jan Huyghen van Linschoten, published in 1596, seems to have caused Prince Maurits van Nassau, Count of Holland, to decide that he wanted such a machine. He therefore asked his tutor and advisor Simon Stevin, the engineer and mathematician, to design him one. This landyacht was used from 1599 onwards on various state occasions. Initially it was used to impress captured Spanish officers with advanced Dutch technology. The vehicle could carry more than 30 people and on at least one occasion reached a speed of over 34 kmh. One smaller companion version was also built and both are thought to have survived until the early nineteenth century. This landyacht was not

Working scale model of a wooden land yacht built by Simon Stevin in 1599 (model built by Paul Nieuwenhuis)

only powered by the wind, it was also built of wood, another renewable resource (although oak was used in this case, which takes a rather long time to renew itself) and must thus be regarded as the ultimate – or perhaps the only – 'green' land vehicle.

Further experiments were carried out with this technology and in 1834, the Frenchman Hayquet designed another landyacht, using three masts and ten sails. Wind-powered vehicles have severe limitations, however. These vehicles could only really be used when the wind was coming from behind, and it is known that Prince Maurits' vehicle had to be towed back by horses after a successful outing. The wind also had to be fairly strong to move these relatively heavy vehicles. It is not surprising, therefore, that landyachts are only really used as a leisure object nowadays. Some wind-assisted designs have been seen recently and at least one of these was run in the Solar Challenge held in Australia in 1988.

External combustion

Steam engines

After the wind, steam is possibly the oldest source of power for self-propelled vehicles. Steam is also interesting because it represents the first motive power generated through combustion. Steam formed the basis of the Industrial Revolution and it is fair to say that our modern industrial civilization would not exist if it were not for the fact that our ancestors managed to discover and control the power of steam.

The steam-powered vehicle predates these events by nearly a century. It is believed that the first steam-powered vehicle ran at the Chinese imperial court in the latter half of the seventeenth century. This device was the brainchild of a Flemish Jesuit, Father Ferdinand Verbiest (1623-1688), who spent most of his life as a missionary in China and adviser and astronomer to the imperial court. Verbiest found references to a small steam-powered vehicle in Chinese sources of around 500 BC and tried to recreate such a vehicle. He succeeded in about 1655, although little is known about this vehicle. It probably used a jet of steam from a small boiler to drive a turbine, which provided direct drive to the wheels. The vehicle was used to cart dishes around the imperial dining table. The principle used was derived from the aeolophile invented by Hero of Alexandria (second century BC) and developed by Giovanni Branca (1571-1645), though Verbiest is the first who is known to have applied this principle to a vehicle.

Newton returned to Hero's original principle and developed a jet-propelled steam-powered vehicle in 1680. He predicted that his invention would ultimately lead to the construction of vehicles that would allow travel at speeds of more than 50 mph! Throughout the eighteenth and nineteenth centuries, designers, especially in the UK and France, developed steam-powered vehicles of various kinds and in the UK some of these were even used for regular scheduled passenger services on some routes. Cugnot in France was probably the first to develop a steam-powered vehicle for road use. His 'Fardier' was designed to tow guns to the battlefield and two were built in 1769/70. In Britain, Trevithick, Gurney and others built steam carriages and by the time the petrol-engined car came along, steam technology was well established and trusted by a public by now used to steam trains, ships, agricultural machinery and road vehicles. Some of these can still be seen today at steam rallies, where the showmen's engines are a particularly attractive feature.

From 1885 to 1925 steam cars competed with the petrol engine, albeit with diminishing success. Great names from this period are Stanley, Locomobile, White and Doble of the US, while in Europe such firms as De Dion, Serpollet/Gardner-Serpollet gave up steam earlier. In fact, in the 1920s, only Stanley and Doble could be described as significant producers, Stanley in terms of volume and Doble for performance and price. These cars were probably the most sophisticated steam cars ever produced in series. Some optimists entered the field at this time, such as the American Steam Car Company (1929-31). However, the Wall Street crash killed off most of them while others resorted to converting the occasional petrol car to steam power. Abner Doble worked for Sentinel in the UK, whose last steam-powered truck was listed in its catalogue until 1951, but essentially the steam car was dead.

A few attempts have been made since 1940 to revive the steam car, or reassess its viability in the light of new technology, but the same problems (see below) continue to interfere. Some car manufacturers have been experimenting with steam, among them Saab and General Motors. Both tried this technology as a possible answer to the toughening emissions legislation in California in the late 1960s and early 1970s, while Ford tried a Williams Engine Company steam engine in one of its cars at around the same time. Abner Doble persevered with his work and individuals and companies around the world continue to this day. Some of the more significant names in this respect are those of Singer and Donald Healey (of Austin-Healey fame) in the UK, Keen and Lear in the US and Gvang and Pritchard in Australia.

Bill Lear, of Learjet fame, was a great steam fan and invested large amounts of money in steam car research; several cars were converted, but the project never lived up to expectations and largely died with him. The Australian Gene Van Grecken designed and built his Gvang prototype in the early 1970s and he successfully overcame some of the steam-engine's shortcomings with the use of electronics. He clothed the car in an attractive streamlined aluminium body and although little known today it must rank as one of the most impressive and advanced steamers built to date.

In common with all external combustion engines, the steam engine suffers from an internal inefficiency which is due to the fact that the heat, produced outside the engine, has to be transferred to the boiler. This takes place through a wall and a certain amount of heat is obviously lost in the process. Another problem is that external combustion engines rely on a temperature difference and after the heat has been used, it has to be removed, as fresh steam has to be made from water. For this reason, a condenser is normally incorporated, essentially a kind of radiator. The heat lost can be used to warm up the occupants in winter, but otherwise it is just wasted, as it is in an internal combustion engine.

A steam engine can generate enormous internal pressures and this means that the engine has to be much stronger and often much heavier – especially with the ancillaries needed to generate and control the steam – than a petrol engine. Even with this margin, internal stress can lead to rapid wear and loss of tolerances, which in turn leads to loss of maximum efficiency at a relatively early stage, although a steam engine will actually continue to work for many years. In relation to petrol-engined cars, steam cars were normally heavier (despite the absence of clutch and gearbox) and more expensive. Another problem was the time taken to generate steam; on early Stanleys this took at least 20 minutes, although Doble managed to reduce this to one minute using flash boilers.

Nevertheless, steam is still used, though not for cars. Power stations generate power by means of steam, using coal, oil, gas or nuclear fuel to generate the heat that produces steam. Very large generators at enormous pressures and very high temperatures are used. This is not the sort of thing we would want in our cars, even if we could accommodate it. Furthermore, power stations, like many ships, use steam turbines under very limited operating conditions incompatible with car use (like the gas turbine discussed later). Even the latest steam car designs do not reach the thermal efficiency levels of a petrol engine, although certain of its advantages may make the steam car viable again in a changing world, where perhaps its multi-fuel ability may make it attractive one day. At present this seems unlikely.

The Stirling engine

Another external combustion engine that was considered promising at one time is the Stirling engine and its variants. This works essentially on the principle that a hot gas expands and a cold one contracts. This is used to move a piston in a cylinder one end of which is kept hot and the other cold. The temperature difference leads to problems similar to those experienced with the steam engine, although various improvements have been introduced. Like the steam engine, the fact that the fuel is merely used to produce heat makes the Stirling engine very insensitive to the type of fuel used and this flexibility makes it very attractive for certain applications.

The Dutch electronics giant Philips invested considerable resources in this engine and several experimental buses used to run around its native city of Eindhoven. However, although research continues to be carried out around the world, this engine too is waiting for some major breakthrough or environmental change before it becomes suitable for road vehicles. Some variants are used to convert solar heat to energy; others thresh rice while being powered by the wasted rice husks.

Electricity

Archaeological evidence suggests that the Sumer people of Mesopotamia were familiar with the principle of the battery. What they did with it is not known, but the battery became the key to the electric vehicle. The first known electric car was a small model built by Professor Stratingh in the Dutch town of Groningen in 1835, but the first electric road vehicle was probably made either by Thomas Davenport in the US, or Robert Davidson in Edinburgh in 1842. However, they had to use non-rechargeable electric cells and it was not until the Frenchmen Gaston Planté and Camille Fauré respectively invented (1865) and improved (1881) the storage battery that the electric vehicle became a viable option. By the turn of the century, the electric car was a commercial proposition, especially for use as taxis, and such was the confidence in its suitability for town use that the French observer Hospitalier declared in 1898 that 'the petroleum spirit cab will never be a practical proposition in large towns'. He may yet be proved right.

The one event that gave electric cars the necessary credibility was the world land speed record set in France on 29 April 1899 by the Belgian Camille Jenatzy with his unique streamlined racing car, 'La Jamais Contente', the first car to go faster than 100 kmh. No petrol car could achieve such speeds, but one of the features of the electric vehicle was – and is – that you can either use your power to travel a long distance at a moderate speed, or you can choose, as Jenatzy did, to travel a short distance at high speed.

The electric car was marketed primarily on its simplicity and as with the automatic DAF car much later, this seemed to appeal especially to women. The typical user of an electric car in the first two or three decades of this century was a middle-aged middle-class woman who used her electric car for shopping trips and visiting friends. These people drove at moderate speeds and did not cover very long distances. For this sort of use, it was ideal. An overnight recharge was normally enough to cover the next day's use.

The electric car was particularly popular in the US and companies such as Detroit Electric aimed the product at their female customers by using fully enclosed bodywork with such items as curtains and flower vases! Electric cars were very heavy and most of their power was in fact used to move the great weight of the batteries; on the French Electrolette of 1902,

The Pope Electric of 1907 (photo Paul Nieuwenhuis)

for example, 60% of the weight was accounted for by the batteries alone. For certain types of customer and certain types of use, the electric car provided an excellent solution; simple controls (no clutch or gear change), smooth power delivery and quiet running had an appeal that for many people outweighed the disadvantages of limited range and limited speed. Some companies did try and prove that their electric cars were suitable for more intensive use, but often had to resort to very low speeds to prove a greater range. Alternatively, your car could be recharged on the way, much like a stop for petrol, although it did take much longer, even if this was done by changing the batteries.

Petrol-electric hybrid systems – as opposed to the battery-electric systems described so far – were used to achieve the smoothness of electric power without the need for heavy batteries that had to be recharged. One of the first was the Austrian Lohner-Porsche. Lohner was the first Austrian car manufacturer, starting production in 1896. However, the company soon turned to electric cars and in 1898 engaged the young Ferdinand Porsche. He developed the so-called 'Radnaben' (wheel-hub) system, which employed an electric motor in each front wheel, powered by a generator, driven by a petrol engine. No gearbox was needed and the generator was used as a starter motor for the engine. When going downhill, the electric motors were used as generators to charge the batteries. This process is known as regenerative braking and it greatly extends the range of an electric car. The

Lohner-Porsche system, though elegant, was relatively complex and heavy. The company even built some racing cars using this principle, but it proved too expensive to compete with the petrol engine. After 1905, their main product became trolley buses and car production stopped. Kriéger of Paris also produced petrol-electrics, as well as an experimental gas turbine-electric in 1908.

The electric car enjoyed a brief revival in the 1940s. The ravages of war and Nazi occupation had made petrol a scarce commodity and in this climate the electric vehicle became once again a viable proposition. A number of small and very simple electric cars were built in limited numbers, especially in France.

The immediate postwar period saw little activity on the electric vehicle front. During this period many small, simple and very light vehicles were built, but nearly all of these used small internal combustion engines in a climate where oil-based fuels, though expensive, were nevertheless regarded as well nigh inexhaustible. However, over the next few decades, battery-electric vehicles gradually became more and more popular for specialized uses. From baggage handling at airports and mail transport at railway stations to milk delivery floats and forklift trucks in confined warehouses and even golf buggies, the electric vehicle has become an integral part of society and yet for personal road transport it has remained relatively unpopular.

During the 1960s, a number of people became aware of the increasing problems of urban traffic congestion. The concept of the specialized compact urban vehicle was born. These were small, slow and often used electric power. A number of car manufacturers became involved and presented some experimental designs. A typical example of this period is the Ford Comuta, developed in 1967. This prototype had four 12-volt batteries, which powered the rear wheels via two motors. The Comuta offered seating for two as well as some luggage space. Similar specialized urban vehicles were developed all over Europe, like the Italian Urbania and the Zele, built in series of several hundreds by Milan coachbuilders Zagato. Several petrol cars were also fitted with electric running gear. Such experimental vehicles were based on the Fiat 500 and 850, Mini, Ford Transit, and many other mainstream donor vehicles. In the US, C.H. Waterman Industries of Athol, Massachusetts converted the Dutch DAF 44 and 46 models to battery-electric power, whereby the Variomatic CVT was retained to provide extra speed and range. Lucas developed a petrol-electric hybrid car in 1982.

The problems of the battery-electric remain the same: heavy batteries, limited speed, limited range between recharging and relatively short battery life. A major breakthrough in battery design has been promised since the early years of this century, but despite various minor improvements, the big revolution has not materialized. Fuel cells, which actually generate electricity, were developed for space craft in the 1960s. Some experimental

C.H. Waterman's electric conversion of the DAF 44 (photo Kazmier Wysacki)

vehicles have been fitted with them, but so far this route seems to have led nowhere either.

Modern battery-electric vehicles

By 1980 there were around 120,000 electric vehicles in use worldwide. Of these, more than half were used in the UK! (Think of milk floats). It is not surprising, therefore, that the UK has regularly shown new initiatives in this field. On 10 January 1985, Sir Clive Sinclair, the British inventor, launched his controversial C5 electric tricycle. This single-seater commuter vehicle was intended as a cheap first model to generate funds for the development of a whole range of electric cars. The C5 was introduced at a price of £399+p&p. Unfortunately, it did a lot to discredit the electric vehicle in the UK.

The demise of the C5 once more relegated the battery-electric vehicle to specialized and experimental uses. However, a number of manufacturers and specialist conversion companies offer normal cars that run on batteries. It has been suggested that in Europe around 6 million cars and 1 million local delivery vehicles could profitably be replaced with electric versions. This would cut emissions by 2.5% at 1987 levels. Oil consumption by traffic, however, would be cut by 3.5%; 1% of European oil needs. By the middle of 1987, for example, Volkswagen had already sold 70 of its electric Golf-based CitySTROMer cars with conventional lead-acid batteries.

A number of new battery types have also been developed, among them nickel-cadmium, zinc-bromine and sodium-sulphur. The sodium-sulphur high energy battery has been developed by – among others – ASEA-Brown Boveri (ABB) of Switzerland and it is used in experimental vehicles from Volkswagen and BMW. The major advantages are four times the energy density of lead batteries, therefore lower weight, and a longer life expectancy with complete freedom from maintenance requirements. However, these batteries have to operate at a relatively high temperature of between 299 and 329 degrees centigrade for the chemicals to reach the right levels of viscosity. Some of the energy produced therefore has to be used for heating. On the other hand, this constant operating temperature makes the unit relatively insensitive to variations in ambient temperature, while the cells also have the advantage of being rechargeable at any convenient time. A complete charge takes two hours and allows 90% of the energy to be used. This would give a 1500 kg car a range of 150 to 250 km, more than adequate for urban and suburban use.

Apart from ABB, research into the sodium-sulphur battery is taking place at Chloride in the UK – which has a Bedford van powered in this way – and Yuasa and Hitachi in Japan. Although these batteries still require the production of certain raw materials, there is at present no serious shortage of the chemicals involved, while the electric infrastructure already exists in most developed countries; a normal electric socket can be used to recharge the batteries. One advantage of electric vehicles is that no energy is used while the vehicle is stationery, e.g. at a traffic light, and this is one of the reasons why greater use of electric vehicles leads to a far greater saving in oil than increase in electricity consumption. Besides, a vehicle would typically be recharged at night, when the demand for electricity is lower and off-peak surplus power can be used.

BMW first built an electric car in 1972, based on its O2-Series body, but in the late 1980s fitted some of its 3-Series cars with the Asea-Brown Boveri (ABB) batteries. The car features front wheel drive, a 72 mile range in urban use conditions, a top speed of 100 kmh and a 0-50 kmh acceleration of 12 seconds. These cars are also fitted with a two-speed automatic transmission, which allows different gearing for climbing hills and driving on the level. Electronic control units engage regenerative braking and monitor recharging and various other functions. The Volkswagen Jetta CitySTROMer uses similar batteries and reaches a maximum speed of 105 kmh and a range of 120 km with 0-50 kmh acceleration in 12 seconds.

Mercedes fitted some 307-type vans with electric power and sold some to local authorities. The El-Trans from Denmark is more like the C5 in being a single-seater commuter vehicle, albeit enclosed for better weather protection, with a range of 40 miles and a 25 mph top speed. It does, however, take eight hours to recharge its conventional battery. Production started in 1988 and sales may reach 10,000 a year in Denmark, where special incentives are offered. PSA already has a comprehensive electric vehicle programme. The company offers electric conversions for its Peugeot 205 and Citroën C15 vans. Examples of these are used by Electricité de France, the Belgian electricity board and the French city of La Rochelle.

In Japan, Nissan developed an electric car motor, which allowed an 18% improvement in range over its older model. Nissan's electric vehicle could go for 180 km on one charge. The Japanese Ministry of International Trade and Industry (MITI) is funding a programme for the development of an electric car using nickel-zinc batteries. MITI invested $20.9 million in the project in 1988.

Since motor racing was banned in Switzerland, the Swiss Automobile Club has been organizing the GP-E, the grand prix for battery-electric vehicles, and many experimental designs are pitched against each other at this event. More and more people are trying to prove the viability of the electric car and in 1987, Mr Karl Bowers drove the 125 miles around London's orbital motorway, the M25, in his homebuilt electric at an average speed of 45 mph and an estimated recharge cost of 57p. He repeated this feat in 1988 and this time it won him a place in the *Guinness Book of Records*.

The EC is also constantly updating its policy on electric vehicles and Transport Commissioner Van Miert aims to introduce several million electric cars by the end of the century. It is recognized that only changes in legislation, whether national or at EC level, will bring about such a major change.

The 'Impact'

If a major breakthrough has not yet occurred, the nearest thing to it must be General Motors' new Impact experimental battery-electric sports car. The great innovation of this car is that it has a very compact and efficient electronic unit for converting DC to AC current, thus enabling the use of much more efficient AC motors, while still using conventional batteries to store power. Other advantages are inherent in the fact that the Impact was designed as an electric vehicle, rather than being a car with batteries and motor added as an afterthought. Main features include very low weight due to extensive use of composite materials (to counterbalance the weight of the batteries, fitted in a long tunnel running the length of the car; this has become a convention among US electric vehicle designers), low drag, fast acceleration, high top speed and good range. The Impact's specification includes 32 12V AC-Delco batteries, 2 motors, front-wheel drive, regenerative braking, a top speed of 100 mph, acceleration from 0-60 mph in 8 seconds, a range of 120 miles, and a recharge time of 6 hours.

General Motors intends to begin commercial production of the Impact – or a development of it – by 1994. However, there is still much work to be done on improving the performance of batteries, and to this end the US 'Big Three' (GM, Ford and Chrysler) have announced plans to co-operate in the development of new battery technologies for electric-powered vehicles.

General Motors' 'Impact' electric sports car (photo courtesy of GM Opel)

This announcement came in January 1991, following discussions which also involved the US Department of Energy and the Electric Power Research Institute, which groups together the electricity producers.

Solar-electric vehicles

The viability of solar-powered vehicles came with the development and improvement of the solar or photovoltaic cell, largely as a result of space technology. The photovoltaic cell converts light into electricity (as in solar-powered calculators). An experimental solar filling station was built at the Hannover trade fair in 1988 by a consortium which included AEG and Varta. The 42.5 metre high structure is covered with photovoltaic solar panels. This 'Solar Tankstelle' supplies 15 kW by maximum sunlight and is used to recharge small battery-electric vehicles which run around the fair grounds.

The Swiss now regularly organize the Tour de Sol grand prix for solar-electric vehicles, but the great leap forward in terms of this technology was without doubt the World Solar Challenge first run between the Australian cities of Darwin and Adelaide in November 1987. Several large companies as well as smaller teams of private individuals developed special vehicles for this event. The famous alternative energy team of Jonathan Tennyson from Hawaii entered a solar, but wind-assisted vehicle, while Ford constructed the Model S. The winner in all respects, however, was General Motors' SunRaycer, one of only three out of a total of 22 competitors to reach the finish. This unique vehicle was covered with 7200 solar cells and reached an average speed of around 66 kmh, although its top speed was measured at 113 kmh. GM invested $15 million in the project. Part of this went into the development by GM's Delco Remy division of a new electric motor fitted with Magnequench high-power magnets, which delivers about 40% more power than conventional types.

The World Solar Challenge was held again in November 1990 and this time attracted entries from Nissan, Honda and Mazda, as well as several academic institutions. The Swiss Spirit of Biel II, entered by Biel School of Engineering, won the race. Honda came second, with the University of Michigan in third position. The winning vehicle's average speed was recorded at 65.184 kmh, a little slower than the record set by SunRaycer in 1987. However, this was an impressive performance by the Swiss, who finished the race well ahead of Honda's high-budget entry. This success was due in part to the new type of solar cells used by the winners. On these cells, developed at the University of New South Wales, the surface is significantly enlarged by laser-etched grooves, which enhance efficiency.

Switzerland should currently be considered the home of the electric vehicle in Europe. Many types of electric vehicle are available in Switzerland and bought by customers. Many users have solar panels on their roofs, which recharge batteries during the day to recharge the vehicles overnight.

Return of the hybrid

The electric hybrid vehicle, whereby a combustion engine is used to generate electricity, which is then used to power electric motors to drive the car, dates back to the early 1900s. This type now appears to be making a comeback and it may well be the most viable version of the electric car in the short term. The return of the hybrid is largely due to the fact that air pollution from car exhausts is at its worst in urban centres where many cars belch out poisonous fumes while stationary in traffic jams. The idea is that the electric motor is used in towns and cities; when a certain speed is reached the internal combustion engine cuts in to drive the car and recharge the batteries.

Both Audi and Volkswagen have recently shown prototypes of hybrid versions of their normal cars. The VW Golf Hybrid is very elegant, featuring a very compact electric motor between engine and gearbox with a clutch on either side. This motor can drive the car, recharge the batteries as a generator and start the diesel engine as a starter motor. The hybrid avoids the problems of battery weight, limited speed and limited range associated with the battery-electric vehicle, and although its internal combustion engine still causes pollution (a catalyst is fitted to both prototypes), the level of emissions is very much reduced, both because the internal combustion engine is used only part of the time and because when it is used it runs under relatively favourable conditions.

Internal combustion alternatives

Alcohol

There are a few alternatives which involve the burning of alternative fuels in what is essentially a conventional internal combustion engine. Alcohol engines (ethanol and methanol) are already used in several countries, with Brazil the most advanced. A few years ago, most cars sold in Brazil were designed to run on a mixture (85:15) of alcohol and petrol. The ethanol was produced from sugar cane and was heavily subsidized by the government as it helped reduce the country's reliance on imported oil. The scheme was criticized by many people for growing food for cars in a country where many of the poor were starving.

From an emissions point of view, both ethanol and methanol produce fewer of the toxins associated with petrol engines, although carbon dioxide levels are often higher and formaldehydes are much higher than with petrol. Furthermore, fuel efficiency is reduced. However, if the alcohol is produced from crops, as in Brazil, the carbon dioxide released can be equivalent to that absorbed by the growing crop in the first place, so that there is no net increase in carbon dioxide. Methanol is highly corrosive, thus requiring changes in materials used in fuel systems; it is also more toxic than petrol. Methanol is usually made as a by-product of natural gas or coal production and treatment, while alcohols can also be made from biomass such as various types of organic waste, including most household rubbish.

Hydrogen

Although alcohol has some advantages over petrol in terms of pollution, it can only be an intermediate step, as emissions are still significant. A truly clean fuel seems impossible, but does in fact exist. Hydrogen, when burned (i.e. made to react with oxygen) produces . . . water. This seemingly perfect fuel is unfortunately extremely difficult to handle. It can be produced in several ways, but the most popular is to divide up water into its constituent oxygen and hydrogen molecules; hydrogen produced in this way has harmless oxygen as its by-product. This method does require large amounts of electricity, which has to be generated first. The second problem of hydrogen is in storing it. Three methods are currently used: metal hydrides (i.e. 'bound' to metals); liquid hydrogen (at very low temperatures of −250 degrees centigrade); hydrogen under very high pressure.

All of these methods are at present either too bulky or too dangerous for use in normal cars, although a number of firms, among them Mercedes,

BMW, Peugeot and Volvo, have experimental cars and commercial vehicles running on hydrogen. Tiny amounts of pollutants are still produced due to the small quantities of lubricant that get burned in the process, but otherwise the emissions are clean. It is unlikely that the storage and delivery problems (filling up with the very explosive hydrogen can be tricky) will be solved within the next 10 to 20 years.

LPG

Another alternative fuel that is sometimes mentioned is LPG (liquefied petroleum gas), a form of natural gas. This is already widely used in countries as far apart as South Korea and the Netherlands. It is burned in normal engines with minor changes and scores well against petrol in terms of toxic emissions (or at least petrol engines without a catalyst). A catalysed petrol engine does better and because of petrol's higher calorific value also overtakes the LPG engine in terms of carbon dioxide emissions, as LPG engines use more fuel than petrol engines.

The gas turbine

The gas turbine is really a kind of internal combustion engine, but the principles involved are quite different. Compression and drive are provided by very fast turning turbine wheels. This principle is widely used in aircraft, but also in power generation, as most power stations use very large steam turbines to power their generators, with steam produced by burning fossil fuels or by thermo-nuclear reaction.

Research into gas turbines for cars goes back a long way, but a real leap forward was provided when aircraft engine experience fed back into car engine development. Shortly after the Second World War both Fiat and Chrysler used such links to develop experimental cars powered by gas turbines. Others followed, including Rover, which successfully entered a gas turbine racing car in the Le Mans 24-hour race in 1963. However, Chrysler took this type of engine more seriously, regularly showing improved prototypes with more and more compact power units during the 1950s and 1960s.

A number of problems were encountered with the automotive gas turbine. First of all, the quality of materials and production systems required is very high, compared with the rather crude conventional engine. Another problem is fairly high fuel consumption, partly due to the fact that to run at its maximum efficiency, the turbine has to run at very high rpm.; figures of 50,000 to 100,000 are not unknown. This feature makes it prone to a very bad form of inertia, comparable to what is known as 'turbo-lag' on turbocharged conventional engines; that is, the engine responds very slowly

Fiat experimental gas turbine car in the Carlo Biscaretti di Rufia Museum, Turin (photo Paul Nieuwenhuis)

to changes in driver input (accelerating) and although this has made the gas turbine quite suitable for racing, where a narrow but high rev-range can be used, in normal urban traffic it makes turbine cars rather sluggish and slow to respond.

Several companies are currently known to be working on further improved gas turbines. Improved ceramic materials over the past two decades have opened up new possibilities and both the Japanese Ministry for International Trade and Industry (MITI) and the EC are supporting research programmes in this field. Both Nissan and Toyota are involved in the MITI programme and have shown prototypes, while the European AGATA (Advanced GAs Turbine for Automobiles) programme has the following members: Volvo, Volkswagen, Mercedes-Benz, BMW and PSA (Peugeot-Citroën). Recently, PSA joined forces with Renault and with the support of the French government they are set to pursue a number of research programmes into alternative power sources. One of these involves a gas turbine-electric hybrid and continues the V.E.R.T. (Véhicule Electrique Routier à Turbine) programme. Toyota in Japan has also been experimenting with a gas turbine-electric hybrid car. This solves some of the gas turbine's problems as it is not used to drive the car, but only to generate electricity, for which it is well suited.

In late 1990, General Motors announced a major breakthrough in gas turbine research. Previously the potential fuel efficiency advantages could never be realized because of heat limitations of the materials used. A new type of ceramic material allows much higher temperatures to be used, with fuel economy improvements of 12 to 15% better than a petrol engine, whilst retaining the multi-fuel capability. This may make a gas turbine for cars or trucks far more viable.

A future for the conventional engine?

The modern car has an internal combustion engine working either on the Otto (petrol) principle or on the diesel principle. There are very few exceptions to this rule, and unlike the early years of motoring, the consumer has no real alternative.

The internal combustion engine has enjoyed a remarkable development over the past hundred years. From an extremely crude and inefficient device, it has turned into an amazingly powerful machine of increasing efficiency, despite the fact that a large amount of its energy is still wasted as heat. In fact, when comparing petrol or diesel with other fuels, they are remarkably efficient in terms of power delivered per gallon and ease of handling, and this means that most alternatives are going to bring new problems, even if they are cleaner.

Manufacturers are aware of this and are seeking new ways of making petrol and diesel engines more environmentally competitive. Engines have become significantly more efficient over the past hundred years and this trend is still continuing. Multi-valve cylinder heads, fuel injection and more and more sophisticated electronic engine management systems allow more power and lower fuel consumption from engines than has been possible before. The revival of the two-stroke engine in a cleaner (though still experimental) form would allow reductions in weight and size, while related concepts such as the Australian Orbital engine also allow significantly higher power.

If consumers put enough pressure on them, manufacturers can rapidly develop such concepts which would allow very small, efficient and low-pollution petrol and diesel engines to power very light and long-lived cars to tide us over until a truly clean alternative becomes viable. Prototypes of lightweight, aerodynamic, highly fuel-efficient cars have already been produced (e.g. Renault Vesta I and II, Citroën Activa, Volvo LCP 2000) and some of these are several years old, showing that producers have been preparing for the current problems for some time. Now it is time to act and we, as consumers and voters, must lead the way; car producers and governments will not act without us.

8: *Towards an integrated transport policy*

The motor car has to date provided a level of flexibility and convenience of use unmatched by any other form of transport. These qualities are the *raison d'être* of the car and account for its phenomenal success. It is unlikely that any transport system in which dependence on the motor car was reduced, no matter how efficiently it was operated, could provide quite the same degree of theoretical freedom and personal mobility. Nevertheless, there are now good reasons why a remodelled transport system – in which private and public transport work in collaboration – should be considered.

The very mobility enshrined by the motor car is under threat. As the car population and road usage increase, the road infrastructure is being subjected to greater strain. The results are slow-moving traffic and tailbacks in urban areas and on some major axes, such as the M1, M6, A1 and the notorious M25. This has serious economic, environmental and psychological consequences. The answer is not to build more roads; we could end up paving and tarmacking the entire country. If the success of the car is not to become self-defeating, a more rational and integrated transport system will have to be developed during the 1990s.

Motor vehicles in general, and not just the private car, are the root of local pollution problems in many urban areas. However, we should consider chronic road congestion and its associated problems as a consequence of social change. To some extent our problems are a culmination of the separation of the home from the workplace, a trend which has been especially evident in recent years and seems set to continue.

Road congestion was not an unavoidable consequence of this social change. Different government attitudes, more coherent transport planning and the development of a more competitive and appealing public transport system, could have given us a different legacy. But this would have had to have been backed up by a public opinion that did not firmly believe in the central role of the car. Instead, public opinion has given the car almost omnipotent status, demanding for it universal access. A change in attitude is only just beginning to manifest itself.

The problem is both philosophical and political. A more integrated and planned transport system would have to define and therefore restrict the role of the car. This would entail a degree of regulation which does not appeal to the liberal values that prevail in this country today. There would have to be a change in government thinking, backed up by a change in public opinion, about the role of the motor car. Such changes in attitude will not take place overnight, although some motor industry executives have publicly expressed the view that the car will disappear from town centres. An integrated transport system, defined as such, does not appear to be at the heart of the transport policy of any of the major parties.

The price of motoring freedom

When we consider the history of Western civilization over the past 10,000 years or so, each major social revolution has brought with it a reduction in freedom. The agricultural revolution which swept through Europe between about 8000 and 4000 BC forced the hunter-gatherer communities to adopt a more static lifestyle. Instead of roaming through nature looking for food, rarely meeting members of other communities, they had to become farmers and stay on the land they cultivated most, if not all, of the time. This allowed a significant increase in population, as the growing of crops and keeping of animals allowed more food to be produced with less effort. However, people became confined to a smaller area and could not move about as freely as before; their freedom was reduced, which allowed a larger community to survive.

Agriculture became more and more productive over the subsequent millennia, as more land was cultivated and agricultural practices improved; each time the population grew to eat the extra food produced. The Industrial Revolution, combined with major improvements in agriculture, allowed a further growth in population, but also a further reduction in freedom; many people now had to live in cramped conditions in rapidly growing industrial cities, but once again, a larger community could be sustained.

The world population has never grown faster than in the past few decades. Despite such schemes as the Common Agricultural Policy, the production of food for this vast population seems to take a relatively low priority. Nevertheless, climate change has already been affecting agricultural production in many parts of the world and if it worsens, other areas, including our own, may well become affected. The only answer seems to be another period of major change. Perhaps in this post-industrial era we see the beginnings of a new environmental revolution, a sudden realization that we have got our priorities wrong and that if we as a species and our planet are going to survive, we had better do something soon. As part of this revolution, as in the previous ones, we may well have to hand in another one of our freedoms for the common good; the freedom to get into our car at any time and drive wherever we like.

We have already lost many of the freedoms associated with the car. The freedom to park where we like has largely gone from urban centres, while traffic lights, pedestrian zones, speed limits etc. have all reduced our freedom as motorists. Although there are still parts of the country where one can enjoy the pleasures of driving, such areas are being reduced, and is sitting in a traffic jam really what motoring is all about? Does this really fulfil the dream of the open road, the delights of the responsive engine and well-handling chassis? And yet, it is this unenjoyable side of motoring, sitting in urban traffic jams, that is also the most polluting and the most wasteful of energy.

Finally, it should be remembered that by losing some freedom as motorists, we gain freedom as pedestrians, cyclists and users of public transport. This trade-off is well illustrated by the Zurich Model described at the end of this chapter. This relies on exactly the principle of restricting the motorist whilst providing more for public transport, cyclists and pedestrians.

Prometheus

The Prometheus programme is the European motor industry's answer to the problems of traffic congestion. Instead of relieving the problem by reducing the amount of traffic using the roads, this project fits roads and vehicles with electronic systems to enable the same roads to accommodate a larger volume of traffic with less congestion, greater safety and greater efficiency.

The programme supports and coordinates the development by the participants, which include most EC governments and vehicle manufacturers, as well as a number of suppliers of various communications and information systems. These systems will inform the driver about the condition and behaviour of his or her car, and about its relationship with its immediate environment. It also envisages systems that allow communication among vehicles on the same stretch of road and between vehicles and the outside world. The main parameters within which all this takes place are the following:

- Safety (accident prevention systems)
- Economy (systems enabling a reduction in energy consumption)
- Environment (reduce the car's environmental impact)
- Efficiency (smoother traffic flow and control for more efficient road use)
- Comfort (increased driver comfort to reduce stress)

These aims are to be achieved by means of (Prometheus terminology used throughout):

1 Improved information for the driver about: obstacle detection (to avoid collisions), monitoring the road environment (e.g. frost detection), driver monitoring (influence on driver of alcohol, drugs or fatigue), vehicle monitoring (road behaviour and vehicle condition/maintenance), vision enhancement (e.g. guidance through fog).

2 Direct driver support: safety margin determination (for prevailing road and vehicle conditions), critical course determination (are we on a collision course?), dynamic vehicle control, supportive driver information (may include the ability to override the driver in order to avoid accidents).

3 Co-operative driving: intelligent manoeuvring and control (e.g. control of overtaking by means of communication between vehicles), intelligent cruise control (smooth control of speed of all cars using one lane), intelligent intersection control, medium range pre-information (information about traffic and road conditions ahead), emergency warning (if an accident happens emergency services are automatically alerted and traffic controlled).

4 Traffic management: static route guidance (route planning), dynamic route guidance (changes along the way), trip planning (shortest route), network control (traffic flow monitoring and control), parking management (guiding the car to a car park or park-and-ride point), flow control (traffic flow per road section, control of overhead lane closure/speed signs, etc.), demand management (e.g. for road pricing), public transport interface (priority to and information about public transport), vehicle fleet management (control of road haulage fleets).

These systems are implemented by means of various interlocking electronic and telecommunications networks and the first results have already been seen on various experimental vehicles. Volkswagen, for example, has shown cars driving closely behind each other at speed without driver involvement. Such a system would allow more efficient use of road space and reduction of driver fatigue and traffic accidents. Various route planning and vehicle guidance systems have also been shown. Such demonstrations are an important part of the programme as they allow both the public and participants to see what progress has been made, and it forces the different participants to co-operate on one project, thus coordinating their activities.

Apart from Prometheus, a number of other projects of varying scope have been proposed or are under way, in Europe and elsewhere. In Berlin, field trials have been carried out on vehicle guidance and information systems, involving 900 vehicles, 4500 intersections, and 1300 traffic lights. In Japan there is the Advanced Mobile Traffic Information and Communication System (AMTICS), while in the US the Intelligent Vehicle Highway System (IVHS) is being developed. This latter programme is one of the largest, and its aims are similar to those of Prometheus, namely to bring about safer, more economical, energy-efficient and environmentally sound road networks. All these programmes involve the creation of 'electronic roads', which control the traffic on them. When combined with electronically controlled vehicles ('drive by wire'), the driver's role in the control of his or her vehicle is gradually reduced, to be taken over by electronic traffic control systems.

Although these systems may well solve a number of problems, they do so at enormous cost in terms of infrastructure and of vehicle complexity and weight, while other problems remain. Prometheus and similar projects may help postpone certain vital decisions, rather than providing a real solution to the problems that face us today. Intelligent roads are bound to come, but economic considerations must largely limit them to urban and major trunk roads in the developed world, and their impact on world transport as a whole will at best be limited.

End-user approach

We have to consider more carefully our transport needs; what do we actually need transport for? And this includes both public and private transport. Given the different means to transport people and goods, which of them are essential for our basic needs, which are essential for our leisure, etc.? Rather than building more and more roads and making more and more cars, we have to consider our actual transport needs.

We need transport to go to work/school, to get our food, to visit friends and relatives, to bring home heavier items such as consumer durables, DIY materials, etc. and we need it for our leisure activities; to go to the beach on a hot weekend, go on holiday etc. For each of these needs we have a choice of transport modes – some are more suitable for some activities, while others are more suitable for others.

The car has really become a multi-purpose vehicle and we must consider whether it is really ideally suited to such a wide range of distance categories and uses. Many multiple car households have already discovered this and many use a small car for local shopping trips and a larger car for longer distances. The car allows us to carry more of our shopping and allows us to get closer to the shops than most public transport systems, which is a great advantage. However, some of these tasks could be carried out in a different way. Many shops already operate a home delivery service for heavier items, and if we only carried the lighter ones home with us while the rest was delivered soon after, a large amount of short distance car use could be replaced by public transport.

The environmental impact of the car is never greater than when it is used for commuting. Each morning and evening millions of cars sit in traffic jams, most of them with one person inside. The car was not invented for this, and if we stop this practice, we can drastically reduce pollution, fuel waste and traffic congestion.

Leisure use of the car, such as visiting friends in the evening, going to the seaside at weekends and going on holiday, is often a relatively efficient use. Relatively long distances are travelled and although congestion does occur, it is limited to certain times of the day or year and the car provides what it does best; a flexible transport system, carrying a number of people and their luggage in relative comfort.

We have, then, isolated three main uses of the car:

1 Shopping; transporting provisions for the household
2 Commuting; going to work/school
3 Leisure activities

The greatest flexibility is required for 3 and here the car performs best. If we could provide attractive alternatives for **1** and **2**, we would greatly reduce the environmental impact of the car. The problems would not be solved, but by limiting car use for **1** and **2** to an absolute minimum and making the remainder as efficient as possible (e.g. car-pooling), we could buy some time for finding more long-term solutions to our problems.

Alternatives to **1** and **2** obviously include public transport, although we have to consider that public transport systems in many large cities are already stretched to their capacity during the rush hours. Demand for these systems may therefore have to be changed as well, by such means as staggered working hours – perhaps reduced working hours to accommodate this – and more people working from home. These trends have already been predicted, but it is quite possible that the very pressure on our transport infrastructure will force us to implement them sooner rather than later. The revolutions in microchip technology and telecommunications mean that there is potentially less need for people to funnel into city centres every day of the week and fan back out again in the evenings. Ultimately it ought to be possible for a majority of office-based workers to work at home at least some of the time. The decentralized office may still be some time off, but the basic technology already exists to make it feasible.

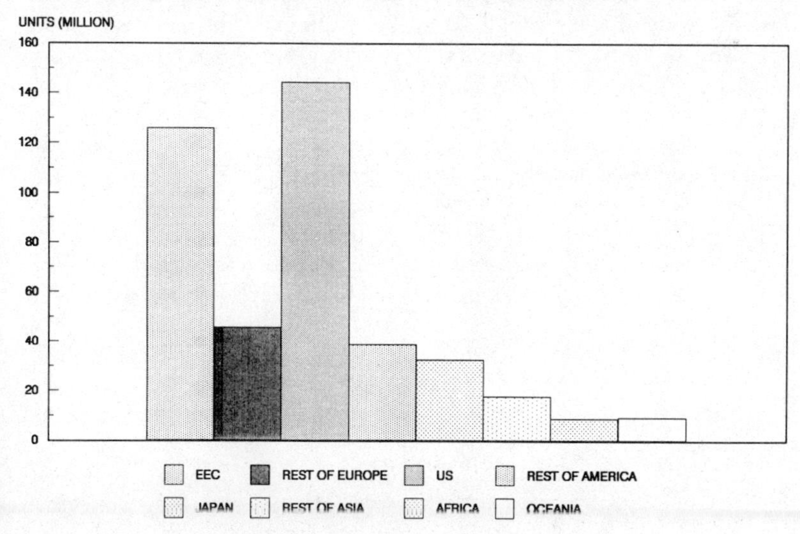

World car population by region, 1989

It is also important for people to reduce the distances they travel to get to shops and to work. Cheaper petrol, company cars and a good car infrastructure have caused people to move further away from their place of work, while employers have chosen locations without considering the effects on the transport infrastructure. Governments should take action to reverse these trends, if our infrastructure is not to grind to a halt. Where people have to travel long distances, adequate and attractive public transport systems will have to be provided. Large employers could play a role here, by sponsoring prestigious and imaginative public transport systems to benefit their own employees as well as other travellers. People who live closer to work could use the bike, and better and safer facilities should be provided for this, while car-pooling could be developed as it has been in the US (see below).

The real cost of the car

We in the industrialized world may think that life is not really possible without the car. In fact, more passenger journeys are carried out by bicycle than by car, and world bicycle production in 1987 was three times higher than world car production at 99 million units compared to 33 million (Worldwatch Institute, 1989). In Asia alone, bicycles transport more people than all the world's cars put together! The bicycle too is a form of private transport, with many of the advantages of the car in terms of flexibility over short to medium distances; and on average it requires only 2% of the capital necessary to own and operate compared to a car.

Less than 1% of the population of the Third World can afford a car, despite the fact that many of their cities are dominated by impressive main roads and multi-storey car parks. Unfortunately, our love affair with the car has inspired many developing countries to motorize, at the expense of their existing low-tech, but not necessarily inefficient, transport infrastructure. The resulting dependence on imported oil often cripples the already strained economies of these countries. In a country like Brazil, for example, the value of oil imports in 1985 was some 43% of imports as a whole.

Despite the advantages of motorized transport, the Confederation of British Industry (CBI) estimates that present traffic congestion problems could cost the UK economy as much as $24 billion each year, and as we have seen, this problem is getting worse. Apart from pollution from the car, the oil industry itself and the need to transport oil over long distances also causes severe environmental hazards. Our need for oil also puts the industrialized countries at the mercy of unstable dictatorships and increases the risk of war, as was demonstrated in the Gulf in 1990-91.

Transport used up 63% of all oil consumed in the US and around 44% of that used in Western Europe in 1985 alone. The price we pay for our petrol does not really reflect the true cost to society, for apart from ignoring the environmental costs, the increased risk of war and the way in which it increases the world's North-South divide, its finite nature is not taken into account. If we change our calculations on this basis, certain expensive alternative fuels may well suddenly look more realistic.

The world car parc of around 500 million vehicles also requires an enormous and expensive infrastructure in order to operate. The US-based Worldwatch Institute in one of its reports, *Rethinking the Role of the Automobile*, reveals that:

> Parking a car at home, the office, and the shopping mall requires on average 4000 square feet of asphalt. Over 60,000 square miles of land in the US have been paved over: that works out to about 2 percent of the total surface area, and to 10 percent of all arable land. Worldwide, at least a third of an average city's land is devoted to roads, parking lots, and other elements of a car infrastructure. In American cities, close to half of all the urban space goes to accommodate the automobile; in Los Angeles, the figure reaches two-thirds.

The car has allowed us to move our facilities over a larger area, thus forcing us to travel longer distances. Those of us who live and work near the centre of a compact town or city, such as York or Norwich, are among the lucky few who do not depend on motorized transport – public or private – to go about our daily business; many in the industrialized world are not so lucky, and although in limited numbers the car provides freedom, in a densely motorized environment it enslaves its users. Many planned cities were designed with the car in mind and although congestion is not always a problem, car ownership is almost a must.

Road haulage and the truck

We concentrate in this book on the private car, but we must mention vans and trucks, because they are significant road users. Our economies rely very heavily on the transport of goods, 82% of which are carried by road; in the case of food this figure is even higher. Commercial vehicles also provide 80% of all public transport journeys, while they are essential for the emergency services and other essential aspects of our society.

Although many of the environmental problems of HGVs are shared with cars, they are often tackled through separate legislation and current thinking concentrates such efforts on emissions and noise. However, a more complete picture should incorporate the following:

- Emissions
- Noise
- Fuel consumption
- Congestion

Emissions

Trucks, like cars, pollute. It is estimated that UK commercial vehicles are responsible for 10% of nitrogen oxide emissions, 3% of carbon monoxide, 3% of carbon dioxide, 3% of sulphur dioxide and 1% of hydrocarbon emissions. Most of them have diesel engines and produce pollutants such as nitrogen oxides, hydrocarbons, carbon monoxide and carbon dioxide, which are also found in petrol emissions, and in addition sulphur dioxide and particles (see Chapter 2). However, trucks do not emit lead or benzene, which is still a problem of unleaded petrol. Diesel is relatively clean and all trucks use it.

The commercial vehicle industry is preparing cleaner trucks and several companies, notably Mercedes, Volvo and DAF were among the first in Europe to show 'clean' trucks. These vehicles are made clean by fitting them with a particulate filter, which still has a number of disadvantages. The filters need to be heated up regularly to burn off the accumulated materials; this carries with it potential risks. A simple oxidation catalyst which reduces carbon monoxide and hydrocarbon emissions is already available on some diesel-engined cars and trucks. The problem of these is that high sulphur content in diesel oil causes the catalyst to emit sulphates as well, which contribute to acid rain. Only the oil companies can solve this one.

On the other hand, many improvements can still be made by cleaning up the combustion process itself. Truck diesel engines still lag behind car petrol engines in terms of engine management systems and this is an area where much development work is taking place. The widespread use of turbocharging and intercooling has not only made engines more economical, but this improved efficiency has also led to cleaner emissions, while relatively simple techniques such as exhaust gas recirculation (EGR) can also be of great benefit. EGR reduces emissions of nitrogen oxides and can be used on diesel and petrol engines. Under this system, some of the exhaust gases are sent round again into the combustion chamber, thereby reducing the combustion temperature and thus the formation of nitrogen oxides. In the fight against emissions, truck diesels are likely to follow developments in car petrol engines. Overhead camshafts, multi-valve cylinder heads and electronic engine management systems are all part of the engine manufacturers' armoury.

Some experiments are taking place with alternative fuels and engines. Several manufacturers have experimental electric vans, while some offer them for sale. Mercedes is experimenting with hydrogen power in vans, while M.A.N. produces trucks that run on natural gas. This M.A.N. engine produces negligible particulate emissions and far less nitrogen oxide than the best diesel engines and it complies with the tough US regulations for 1994. Gas turbines are also considered a promising alternative for HGVs and in Japan government and industry are working on a joint research programme in this field; experimental gas turbine powered buses and trucks have been shown by Mitsubishi, Nissan and Toyota. The pressure of impending legislation, especially in the US, will speed up such developments.

Noise

Another problem is noise, which led the government of Austria to ban trucks from its roads during the night at the beginning of 1990. Four principal sources of noise have been isolated, namely engine, transmission, aerodynamic noise, and tyres.

The industry's response has been to encapsulate the engine and gearbox in sound-insulating materials, although quieter truck diesels are also being developed. Aerodynamic noise refers to the noise produced by the vehicle moving at speed through the air. Aerodynamic aids such as spoilers and fairings are increasingly being fitted as aids to fuel efficiency, and these also reduce noise. As a result, the noisiest part of a truck is its tyres, and research is taking place to reduce this problem too, but both tyre design and road surface have to be tackled. EC legislation has been controlling truck noise for some time and standards are getting tougher all the time. Between 1977 and 1988/89 (directive 84/424), the noise limits changed as shown in the following table.

EC noise limits for commercial vehicles
(All figures in dB(A). GVW = gross vehicle weight)

Category	*1977*	*1980*	*1984**	*1988*	*1989*
Up to 2 tonnes GVW	—	—	—	78	78
Up to 3.5 tonnes GVW	84	82	82	79	79
Over 3.5 tonnes GVW				under 75kw/100bhp	81
Up to 150kw/200bhp	89	87	83	over 75kw/100bhp	83
Over 150kw/200bhp	91	88	85	—	84

*from 1984 the measuring methods were changed, leading to a de facto tightening of standards by another 3 to 4 dB(A).
(*Source: CEC, Febiac*).

In view of the nature of the decibel scale, these are very significant improvements (a one point reduction amounts to a tenfold reduction in noise).

Fuel consumption

In one respect in particular, commercial vehicles have for some time been more environment-friendly than cars. Fuel efficiency has been a major consideration among commercial vehicle customers for many years, as fuel costs form a large slice of total operating costs in the road haulage industry. As a result, manufacturers have been putting considerable efforts into reducing consumption. On average, trucks use some 35% less fuel than they did 15 years ago, despite significant performance gains. It is quite surprising that a fully loaded top-of-the-range articulated lorry of 38 tonnes GVW (= gross vehicle weight, i.e. the weight of vehicle + load) can return around 8 mpg, only slightly worse than some two-tonne luxury cars carrying at most 5 people and their luggage. Many 17 tonne rigids (non-articulated trucks) can match the two-tonne luxury car in terms of fuel consumption, but with much greater fuel efficiency in view of the load carried.

Despite this remarkable fuel efficiency, further improvements are possible. Tests have shown significant improvements in fuel consumption for vehicles fitted with aerodynamic aids. Relatively simple add-on parts can reduce drag from around .75 to around .4 (matching many passenger cars) and this has led to fuel savings of up to 20%. Scania was one of the first to introduce an improved series production cab specifically to improve aerodynamics by reducing drag by 15%. Engine improvements as outlined above will also improve fuel efficiency significantly.

Another factor in reducing consumption is the driver. Driver training is considered increasingly important, for the most fuel-efficient trucks still need to be driven properly for the full benefits to be realized. The wave of very powerful trucks since the late 1980s has made this even more pressing. The high power outputs are intended to increase average speed by making it easier to climb hills, for example. Driver effort is also reduced, but the driver can use the extra power to drive faster generally. This has prompted some operators to fit speed governors to their trucks, limiting them to 70 mph, and this has led to significant improvements in fuel consumption. Governments and the EC are making such devices mandatory for reasons of fuel efficiency and also safety. This last aim may not be realized as a speed governor would still allow a truck to drive through a village street at 60 or 70 mph.

Congestion

Trucks are very much heavier than cars, especially when loaded, and this can cause greater damage to roads and bridges. Heavy goods vehicles are said to be largely responsible for the need to strengthen the Severn Bridge at a cost of £33 million. Designs with more axles could alleviate this problem to some extent by spreading the load over a greater number of wheels and a greater road surface. The increasing use of air suspension should also help to alleviate the problem.

To move the same amount of goods by lighter vehicles would lead to a very rapid increase in the number of vehicles on the road, as well as an increase in overall pollution and fuel consumption. The following figures illustrate this point; it takes 2.4 17-tonne trucks, 6.9 7.5-tonne light trucks,

24 one-tonne panel vans or 48 car-derived vans to transport the same amount of goods as one 38-tonne articulated lorry. The small car-derived vans would take up 78 times as much road space as the artic, use seven times more fuel at a cost eight times higher to the consumer (figures from *Commercial Motor*, 10/90). In Japan the great reliance on light commercial vehicles is a major contributory factor to the serious congestion problems in urban areas. In the UK, the introduction of the 38-tonne truck has actually led to a reduction in the total number of trucks on the roads by some 12,000.

Nevertheless, future developments may see a change in our reliance on road transport. Longer distances can often be covered more efficiently by trains or ships, although these too cause pollution and use fuel. Road vehicles could concentrate on local distribution or the serving of areas away from the railway network, or for jobs where greater flexibility is needed. Flexible container systems could facilitate the interface between road and rail for these purposes. Countries such as Germany have a deliberate policy of limiting road transport and making the railways more attractive for the transport of goods by means of subsidies. All German car plants, for example, have a rail link, used both for the supply of parts and for the dispatch of finished cars. German industry enjoys relatively cheap transport costs in this system.

However, not all goods traffic could be carried by our existing rail network, nor even an expanded and heavily subsidized network, as in Germany. Road haulage will therefore continue to play an important role in our future traffic systems, especially as it offers a unique flexibility. The transport of goods vital to our economy should really enjoy a higher priority than the transport of large numbers of commuters, each in a car, and the reduction of car commuting would make our road haulage system more efficient and less wasteful of resources, as journey times would be reduced. Nevertheless, great efforts should be made to make road transport as environmentally efficient as possible, and alternatives should be used wherever practicable. As in an integrated transport system (see below), different modes of goods transport should be made to co-operate rather than compete, and changeover points should be as efficient as possible.

An integrated transport system?

An integrated transport system is one whereby the different modes of transport are not in direct competition, but instead co-operate to produce as far as possible a door-to-door transport service. The changeover between one mode of transport and another (the interface) is crucial in such a system. In fact, such systems already exist in an informal way. The London commuter, for example, may drive to his/her suburban railway station by car, get on the train, change over to the London Underground and finish the journey on foot. In this way, four different modes of transport are used to complete one journey in what is, under the circumstances, a relatively efficient manner. However, in this system, active co-operation of one mode of transport with the requirements of the others is very limited. Apart from, perhaps, providing parking facilities at the station, and having train and tube stations in the same place, little is done to facilitate the changeover from one mode to another.

The system has already been taken one step further on those journeys where British Rail carries bicycles for free. Alternatively, there are train routes where cars can be carried over long distances (or through tunnels). The advantage of these systems is that the flexible private transport systems (bike or car) are available at each end, for the most individual part of the journey, while the public transport system is used over a standard route where it is at its most efficient. Nevertheless, most transport systems still consider themselves in competition with each other, and such sentiments are encouraged by government policy. New priorities will have to guide transport systems. Environmental considerations will have to be taken into account, and these may often override such criteria as profitability. Nevertheless, a popular, attractive public transport system may well be profitable as well.

Changeover points in an integrated transport system are of two basic kinds; those between public and private modes of transport and those between two types of public transport. Private transport will mostly be used for the early and later stages of a journey, where individual requirements are greatest. This is already often possible where the traveller lives within easy walking distance of a public transport interface (bus stop, railway station, tube station, etc.) and where the final destination is again within easy walking distance of a public transport interface. Where the distances to the nearest public transport interface are longer, some sort of vehicle may be used. Bicycle use would be more attractive if all types of public transport were designed to carry bikes, because then this form of private transport would also be available at the other end of the journey. Bicycle hire or 'white' bicycles (free, publicly owned) at the public transport interface nearest the traveller's destination would also be a possible answer.

The car presents greater problems. In many cases it can be used to connect with public transport in the initial stage of the journey (for instance, driving to the station and leaving the car there), but it is rarely available at the other end, where it may often be unsuitable anyway (e.g. in a city centre). The Witkar (see below) would be a possible solution. One could even envisage an extension of the system whereby one could put a car on the train, thus having it available at the other end, but this system would be more suitable for longer holiday journeys than for daily commuting, at least with present-day cars.

Further improvements could also be introduced in the interface between public transport systems. Convenient high speed rail links with major airports could relieve congestion around these areas. Train fares will have to come down to be competitive with car travel. Timetables of trains, buses, underground railways, etc. could be better co-ordinated, while additional systems such as monorails and trams (or light railways) could be used in urban areas to relieve the bus and provide faster and more direct connections between the city centre and the suburbs. A light railway could run along each of the main axes into a town or city with automatic priority over private transport (cf. the Zurich Model, below) and with stops fed by local minibus lines. Comprehensive seating and some luggage capacity to carry shopping and pushchairs should be provided.

Public transport systems also pollute and can also cause congestion, but their impact is significantly less than that of cars. It has been calculated in the US that a car with one occupant uses up to 1860 calories per passenger mile, a bus 920 and a train 885; so each of the latter is about twice as efficient as the car plus driver (walking uses up 100 and cycling 35(!) derived in both cases from renewable food sources). However, if we look at the question of congestion, we see that public transport systems are even more efficient, as is illustrated by the following table from Worldwatch Paper 90:

Number of persons/hour that 1 metre-width of route can carry

Mode	*Speed (kmh)*	*Persons*
Car in mixed traffic	15-25	120-220
Car on motorway	60-70	750
Bicycle	10-14	1,500
Bus in mixed traffic	10-15	2,700
Pedestrian	4	3,600
Suburban railway	45	4,000
Bus in bus lane	35-45	5,200
Surface rapid rail	35	9,000

Finally, we offer three examples of schemes – already in place, or used in the recent past – which point in the right direction and show what can be done without radically altering our lifestyle.

The Witkar

This project deserves a special mention, as it was part of a far-sighted urban transport philosophy. The Witkar was the brainchild of Luud Schimmelpenninck, a prominent member of the Dutch Provo/Kabouter movement of the late 1960s, which had a strong influence on life in Amsterdam at the time and which has to some extent influenced thinking in the Netherlands ever since. After two members of the group were elected to the local council, the white bicycle plan was introduced. Under this scheme, the council owned a number of bicycles, painted white, which anyone could pick up and use free of charge for journeys within the city centre. They

The Witkar

would then be left at the users' destination, where other people could pick them up, and so forth.

The idea was based on the premise of a total car ban in the centre of Amsterdam. People would then use the trams and buses to get around. For more complex journeys the bike was used – white or privately owned – and if these options were unacceptable, there was the final piece of the system, the Witkar, a brilliant concept combining the advantages of both public and private transport.

The Witkar was a small battery-electric two-seater car of a curiously tall design, with some luggage capacity. These vehicles were based at strategically placed recharging stations at key locations in the city. Members of the Witkar co-operative received a key, which allowed access to any Witkar, and these could then be used for a particular journey and after use returned to the nearest recharging station. The Witkars were built by Dutch specialist vehicle builders Cock and Spijkstaal and had a range of between 30 and 40 km, adequate for this type of urban use. The scheme was started in 1974, but lack of co-operation from the authorities meant that the Witkar had to compete with the private car. The proposed traffic ban from the city – which was never implemented – would have led to a more favourable outcome. Despite several attempts to breathe new life into the Witkar, the scheme finally faltered in 1986. It is, however, more relevant than ever, and with the necessary co-operation from local authorities such a scheme provides a feasible alternative to both private and public transport for urban areas around the world.

The Zurich model

Like other major cities, Zurich suffered from traffic congestion problems. One possible solution suggested in the past was the building of an expensive underground railway system, but this was rejected by the people in referenda in 1962 and 1973. The council therefore decided that they had to improve the efficiency of the existing transport system.

From the early 1970s onwards, Zurich has been developing a system of urban traffic management which gives priority to public transport by means of trams and buses. Over 200 causes of public transport delay were analysed and measures were implemented to combat them in order to make public transport as efficient and attractive as possible, given the existing infrastructure.

Some of the main causes of delay were found to be parked cars or cars turning across oncoming traffic, as well as traffic lights and hold-ups caused by minor collisions. To solve these problems, major parts of the city centre were altered by:

- restricting car parking and waiting
- reducing the number of left turns for cars
- introducing separate bus lanes and tram lanes
- introducing a priority system for public transport

The key element of the scheme is a centralized computer system, which keeps track of all public transport vehicles and makes sure that a vehicle is available in the right place at the right time. Two trams and five buses are kept in reserve and staff can order temporary diversions or run a bus on a tram route. If there is a delay, staff can inform passengers at stops by a public address system of the delay and likely duration. Another important element is a separate system whereby sensors in the road know when a bus or tram is approaching a junction controlled by traffic lights. The lights then automatically change to allow the public transport vehicle quick passage and afterwards change back to favour the cars as fast as possible. So, although public transport has priority, the delay to other traffic is kept to a minimum.

At a relatively low cost, the city of Zurich has improved the quality of life by reducing dependence on the motor car, while increasing overall transport efficiency. The benefits have been similar to those derived elsewhere from building an underground system, but at a fraction of the cost and limited disruption to the existing city centre. For all these reasons it has consistently enjoyed the support of a majority of the people and many have been tempted away from commuting by car into efficient trams and buses.

New forms of commuting in the US

Like other industrialized countries, the US faces serious traffic congestion problems in many urban areas. It has been estimated that traffic congestion causes annual delays of some two billion hours and two billion US gallons of wasted fuel. The average commuting time in the Los Angeles area increased by 25 to 50% between 1989 and 1990 and the problems are compounded by a shift away from the normal commuting from suburb-to-city-centre to more and more suburb-to-suburb commuting.

However, many authorities as well as companies have responded quickly, often reintroducing ideas first tried at the time of the energy crises of the 1970s. In California, the South Coast Air Quality Management District obliges companies with 100 or more employees to submit a programme to reduce the number of vehicles used for commuting. The private sector, however, often takes the initiative. It has been found that with a declining labour force, improving the trip to work can often be used to make a job more attractive, thus helping to recruit and retain staff.

The most popular and widespread solution to reducing the number of commuting vehicles is car-pooling, whereby neighbours who travel to the same place travel together. Various non-profit matching agencies have sprung up to enable as many people to travel together as possible. Some companies have bought minibuses to carry even more people in one vehicle, while various other incentives are also used. Many companies also promote the use of public transport or bicycles.

The 3M company in Minnesota introduced 'vanpools' (minibuses for commuter use) in 1972, while the Hughes Aircraft Company has 250 minibuses for use by its 68,000 employees. Many New York employers offer subsidized public transport tickets, while in other parts of the country bicycle racks and showers are provided for cycling employees. Orange County in California is even considering providing its employees with free walking shoes.

Although various different schemes exist, more and more employers opt for one or more of the following, which are rapidly becoming a standard commuting package for US employees:

- preferential parking for pool cars and minibuses
- free or cheap petrol for pool cars and minibuses
- free weekend use of pool minibus by drivers
- subsidized public transport
- on-site information on public transport, cycling etc., as well as pool-matching agencies

- flexible working hours
- 'telecommuting' (i.e. working from home)
- bicycle racks, lockers and showers
- special arrangements for car hire for meetings during the day or going home late.

9: *What car buyers can do now*

Given all that has been discussed so far, what can we, as motorists, actually do to make some contribution to lessening the impact of the car on the environment? To some extent we are helpless, for we are caught up in a society that, in many ways, forces us to be car-dependent, whether we like it or not. On the other hand, many of us also like cars and enjoy driving and we may be reluctant to give it up, even if it is damaging the environment. Nevertheless, many car fans are also environmentally aware and would like to make their motoring less damaging. The only real answer would be to give up the car altogether, and to be really environment-friendly we should only use human, animal or wind power to get about. We should not forget, however, that the access that the motor car provides has also made many city dwellers aware of nature.

Recent research has indicated that if public transport was more attractive, far more people would use it. The Zurich experience (see Chapter 8) bears this out too. It is also likely that if cycling was safer and enjoyed better provision, more people would use a bike. Here is a start; use the existing facilities and campaign for their improvement. Walking is completely free and could be used more for practical purposes. But, back to the car. Given that we need one – and in many rural and isolated areas of the UK people certainly do, as the roads are unsuitable for cycling most of the year and badly served by public transport – what can we do today to minimize its impact?

Choosing a car

Our efforts should start with the choice of a car, whether new or used, or even old. If we buy new, we can already make a significant difference to toxic emissions by buying a car with a catalytic converter. Cats are now available for many cars, from Minis and VW Polos to Volvos and Rolls-Royces. If it costs more, pay the extra; it is worth it.

If you are buying a used car, see if it can run on unleaded petrol, but don't ruin your engine, because doing that would hardly be environment-friendly. Buy a car that is cheap on petrol. We do not all need to drive around in 2CVs (although a modern equivalent in addition to the AX would be welcome), but within the segment in which you (or your company) buys, choose the most fuel-efficient model. You will find that it is not always the slowest. Also find out if a catalyst can be retro-fitted; this is possible on several older models.

If you have a 12-cylinder car consider an eight or a six next time; if you have a six, see if you cannot live with a four (e.g. a turbo four). But most of us use four-cylinder cars anyway, and we might consider a slightly smaller, lighter, more fuel-efficient car next time. When electric cars become available, if you can afford them, buy one; it is the only way to show the industry and the authorities that there is a demand. Try also to keep your car going as long as possible. Choose a car that will last and maintain it well.

All these criteria are not necessarily compatible. A car that is built to last is often fairly heavy, while a light, fuel-efficient model may be flimsy and lacking in build quality. Besides, many catalyst-equipped cars use more fuel than those without. What matters is that we all start choosing a car on criteria different from those we use today. Colour does not matter, but metallics often cause more VOCs to be released than solid colours. Electric windows, sunroofs, central locking, etc. are all wonderful gadgets, but they also add unnecessary weight. Prestige and image are irrelevant when we choose a car, except where that prestige is the result of quality and a long life expectancy (Mercedes, Volvo) or efficiency. Buying a new car because of a new registration letter seems even less relevant.

When choosing a car, look at the weight of your choice compared with its competitors, and consider its fuel consumption. Official government figures are now available for all cars and car dealers should be able to provide them. Also consider life expectancy and find out the manufacturer's policy and practice on recycling. Ask about the drag coefficient. If you prefer a manual gearbox, go for a five-speed if possible; if opting for an automatic, is CVT available? In 1991 this was only the case on the Ford Fiesta CTX, Fiat Uno Selecta, Lancia Y10, Subaru Justy ECVT and Volvo 340 (this last one is no longer made, but is available secondhand). If you enjoy your driving and are considering a sporty model, remember that a light fuel-efficient car such as the Citroën AX GT can be as much fun as many larger and heavier sports cars and sports saloons, at a much lower cost both to you and the environment.

Safety

More and more people use safety as a criterion when choosing a car. Safety comes in two forms according to the industry – 'active' and 'passive'. Active safety refers to the driver's ability to avoid an accident in a particular type of car. Passive safety refers to the car's ability to enable the occupants to survive an accident if it does happen. Here there are still differences and it is the larger, heavier cars like Mercedes, Volvo, BMW, Saab, Jaguar and Rolls-Royce that score best in this respect, judging by accident survival statistics from countries such as Sweden and the US, where such information is collected.

One safety problem is the incompatibility of the different types of transport on the same roads. If a cyclist runs into a pedestrian, the pedestrian normally comes off worst; similarly, if a motorbike hits a bicycle, the cyclist is likely to be worse affected, while if a car hits a motorbike, the biker's chances are considerably worse than those of the car driver. In the same way, when a light car, such as a Mini, is hit by a heavy car, such as a Volvo, the Volvo's occupants stand a better chance of survival than those of the Mini. If, on the other hand, the Volvo is hit by a juggernaut, the truck driver has a better chance than the people in the Volvo.

All this seems obvious, but it is often forgotten that if we allow such a range of disparate vehicles to use the same road infrastructure, this in itself will increase the road casualty toll. If we all drove lightweight cars or HPVs, we would probably not need the safety cage of a Mercedes or Volvo to protect us, at least if goods traffic did not use the same roads as well. As things are, you may decide to protect yourself and your family as much as possible by opting for one of the heavier cars, despite the environmental impact. We should mention that some smaller cars, such as a VW Golf or Volvo 300, do not score badly when hurled at a concrete block at 30 mph, while experience with human-powered vehicles has shown surprising crash-survivability in these very light vehicles.

Using the car

Rather than always getting into your car automatically, assess each time whether the same journey cannot be made in a more environment-friendly way. Could you walk, take the bike or use public transport?

It is particularly important to reduce commuting by car. If you use a company car for this, consider the real cost, rather than your subsidized car use (see below). Companies too can encourage car-pooling, provide their own multi-people transport and review their company car policy; if an employee is worth that much, why not pay a higher salary, rather than buying a car? Shopping trips often involve carrying bulky items or large shopping bags around and the car is ideal for this. Not every shopping trip involves all that much cargo, however, and in these cases, why not walk, take the bike, or use public transport? Also consider whether it is not possible to live nearer to your place of work or to local shops; not only will that make it easier to help protect the environment, it can also save you significant amounts of money. When choosing a place to live, we should start taking these transport criteria into account.

If you do use the car, adopt a relaxed, economical driving style, accelerating slowly and reducing speed slowly; sudden changes in speed waste fuel. Make sure you treat the engine gently, especially when it is warming up; it will last longer that way. If you use a roof rack, take it off when you are not using it as even an empty roof rack can increase fuel consumption by 10-15%. Keep tyres at the right pressure, shock absorbers in good condition, wheels balanced and properly aligned, as all these influence fuel consumption and wear.

The curse of the company car

The UK is virtually unique in the world in having a car market largely dominated by the company car. In 1990, 52% of all new cars sold were company cars. Obviously, the motor industry and those company employees who benefit do not complain. However, the rest of us should. Not only is this subsidized car use environmentally unsound, the low tax penalty on perk cars means that the taxpayer subsidizes company car users. According to Greenpeace this subsidy amounted to £150 per household in 1990.

Of course some people such as telephone engineers, sales reps, AA/RAC patrolmen and others need a car in order to carry out their job. Most company cars in the UK, however, are given as a perk, essentially a lightly taxed part of people's salary, and 70% of company car mileage is accounted for by private journeys. In many cases petrol, maintenance, road tax, etc. are paid for by the employer as well. Not only does this make car use cheaper than public transport, it makes it cheaper even than running a bicycle and thus leads to a massive distortion of the transport market.

Company cars are also generally larger and less fuel efficient than cars bought privately. The Greenpeace study found that company cars on average had engines 190cc larger than privately bought cars. An Australian study* found that company cars were more likely to have airconditioning and automatic transmissions, both of which increase fuel consumption. This same study found that while private buyers generally put fuel consumption near the top of their list of criteria for choosing a car, company car buyers were more interested in comfort, performance and status. In the UK, company cars are generally also fitted with far more options and gadgets than cars in other European countries. These not only add to the weight of the car and therefore affect fuel consumption, they also affect the average price of cars in the UK market, making cars generally more expensive than in other comparable markets. This affects private buyers as well.

That the perk car stimulates car use is illustrated by the fact that company car drivers travel on average some 1700 private miles per year more than private buyers. Company cars also affect safety, as surveys have shown (cf. Potter 1990) that company car drivers are three times more likely to be involved in an accident than private drivers, despite driving 'only'

*L. Knight, (1990), 'An Australian Study of the Impact of Company Cars on Energy Consumption' in: *Australian Road Research*, 20(2), June 1990. Company cars in Australia account for around 35% to 40% of new car sales, according to this study.

twice as many miles. Clearly, if we are going to make our driving greener, the company car will have to go and only the government can bring this about through realistic taxation. In the meantime, if you have a company car, consider its real cost before you make a private journey and if you belong to the 'user-chooser' category, try and choose a model that is as environmentally sound as possible; so go for a catalytic converter rather than airconditioning.

Maintaining the car

Maintenance is important in extending the life of a car. For the bodywork this means regular washing (but not too often; it can waste water), but not with soap or detergent as there is no need; on a well-waxed car (twice a year will do in most parts of he UK) water alone will shift most dirt. At least as much time should be spent hosing down the chassis, wheel arches, etc. as this is most exposed to mud and winter salt. Having those parts steam-cleaned every spring is a good idea. Most cars receive comprehensive rust-proofing and a six-year anti-rust guarantee has become the norm in the industry. But if we are aiming to keep our cars going for at least 20 years, when that guarantee is long expired, more rust preventative care is needed. After about five years, all cars can benefit from a second anti-rust treatment and several companies will carry this out. However, one of the most effective treatments can be done cheaply at home, using a DIY product called Waxoyl.

If we can keep rust at bay, half the battle is won. Other major items, such as engine and transmission, can all be replaced or rebuilt, albeit at a price. Choose a car that has a proven record of engine and transmission longevity, and here again, many larger cars such as Volvo and Mercedes score well, although such classics as the good old Morris Minor should not be dismissed here. Many smaller cars too have a good record, and life expectancy is increasing throughout the industry. Besides, there is always the option of going for an older car whose engine may not last as long as a Mercedes', but is a lot cheaper to replace when it does go. A few manufacturers are beginning to help in keeping older cars on the road in this way. BMW, for example, has reduced the prices of several parts for some of its older cars – subsidized by increases on new car parts – to enable less affluent and often younger enthusiasts to keep their old BMWs on the road.

It is also important to keep the car properly tuned. Every car pollutes the least when it is in a perfect state of tune. It also uses less fuel when properly tuned, so the owner can actually save money as well as reducing emissions. Regular oil changes can help extend engine life, but if it is a DIY job, the oil should be taken to a recycling point for recycling; surprisingly, many DIY mechanics still pour used engine oil down the drain. Not only is this extremely harmful to the environment, it is also against the law. Other items, such as batteries, can also be recycled, so make sure that this is done. Local Friends of the Earth and other environmental organizations often have lists of who collects what for recycling in your area.

The role of the motor industry

We have already mentioned how industry can and should change. It is important for industry to respond rapidly to new consumer requirements such as low weight, low drag, low fuel consumption and longer useful life. The image of the car as projected by its advertizers will have to change. From being a macho prestige object it will have to be much more environmentally efficient, although this does not mean it can no longer be fun. The motor car will have to be redesigned on the basis of these changing priorities; it can be done, as we have seen, and prototypes exist in the store rooms of many manufacturers' product development departments. If the car industry does not respond, other, more flexible and imaginative companies and public bodies may have to do it for them.

Government policy

Governments should respond rapidly to changes in public opinion and should also themselves shape public opinion and push it into a more environmental direction. Government and other public bodies could change their vehicle buying policies along more environment-friendly lines.

Government can also support research into alternative fuels and alternative car concepts. Such support already exists in Japan, Germany and also France, where the state has invested in a joint development project led by Peugeot S.A. and Renault, which covers a range of possible alternatives (see Chapter 7). The EC also has a role to play in this. Governments may also have to give incentives for cars using alternative power sources and should publicize these where they exist; for instance, there is no road tax for electric cars in the UK, but how many people know this?

If we don't support the development of greener technologies, our competitors will rapidly overtake us. It may be necessary to support entities other than vehicle producers for the development of greener cars. Academic and research institutes, small innovative businesses – whose thinking is not hampered by the fact that hundreds of millions of pounds have just been invested in production systems geared more to today's cars than to those of tomorrow – all these have a contribution to make.

Tax structures will before long reflect the fuel consumption of a car, rather than a fiscal horsepower rating (France, Spain), weight (Netherlands) or no differentiation at all (UK). Italy uses a car taxation system which approaches this idea in that it penalizes any car with an engine over 2 litres capacity (petrol) or 2.5 litres (diesel). This system has in no way depressed the Italian market, nor has it made its local producers less profitable; it even means that in Italy you can buy a 2 litre Ferrari, not available elsewhere. Japan has a similar 2 litre tax break. A more gradual system with major tax breaks at 1 litre, 1.5 litres, 2 litres, 2.5 litres, 3 litres, etc. may allow a better control of the market, and such a system could be used to provide incentives for the purchase and use of cars of less than 1 litre capacity. For many years, Japan has been supporting its microcar segment in such a way.

Voters may also want to consider the environmental policies of the different parties, especially as they affect traffic. Electing a government that has a stated commitment to radical reform in this respect would help speed up change in a more environment-friendly direction.

The Dutch environmental policy plan

The Netherlands is the most densely populated country in Europe. Traffic problems often arise there before they do in other countries, and this led in early 1989 to the publication of the most radical environmental white paper presented by any European government.

The Dutch National Environmental Policy Plan, entitled 'To Choose or to Lose' in its English translation, was directly responsible for the fall of the coalition government and as a possible precursor of policies elsewhere in Europe it is worth a closer look. Although it deals with environmental issues generally, the sections dealing with transport are the most relevant here, so we will concentrate on those.

The NEPP was written in response to government forecasts which predicted an increase in car use from 75 billion vehicle kilometres in 1986 to 120 billion in 2010, an increase in the number of cars on Dutch roads from around 5 million in 1986 to 7 or 8 million by 2010. In addition, goods traffic by road would see an increase of 70% to 80%, although as a result of emissions legislation, pollution per vehicle would drop by 80% to 90% for cars and 50% for trucks.

The main objectives of the NEPP in response to these forecasts are as follows:

- 'Vehicles should be as clean, quiet, economical and safe as possible and . . . made of parts and materials which are optimally suited for re-use'.
- 'The choice of mode for passenger transport should result in the lowest possible energy consumption and the least possible pollution. On the basis of anticipated technical developments this means a preference for public transport and bicycles in the coming decades. Great attention is also being paid to reducing energy consumption and environmental pollution in goods transport'.

- 'The locations where people live, work, shop and spend their leisure time will be co-ordinated in such a way that the need to travel is minimal'.

As well as setting a number of targets for the reduction of emissions between 1986 and 2010, the plan envisages increased spending on public transport and bicycle facilities, while the price mechanism will be used to encourage one mode of transport over another (i.e. car use will be penalized). These measures will be used to bring about various shifts from one mode of transport to another:

- over short distances (5-10 km) from car to bicycle
- from car to public transport to double the number of passenger kilometres covered by public transport
- for longer distances of up to 1000 km a shift from air to rail

Car use will be actively discouraged by means of increasing variable costs such as increased excise duty on fuel. Commuting by car will also be discouraged by removing subsidies and encouraging car pooling, private group transport (e.g. minibuses provided by employers) and public transport alternatives. Private companies will be asked to draw up kilometre reduction plans. To facilitate the use of a range of public transport systems, fares will be integrated so that a single ticket is bought for a journey rather than a separate ticket for each separate transport mode used during that journey (one step towards an integrated transport system). Co-operation between transport regions and systems will be actively encouraged. Planning policy will be used to discourage labour-intensive and visitor-intensive businesses from locating where they cannot easily be reached by public transport.

An updated version of the NEPP was released in May 1990, but some of the radicalism had by then been lost through compromise. The original stands, however, as a fully costed and realistic example of what could be achieved if the political will existed. Only public opinion can now create this political will.

In 1990, the UK government produced its white paper on the environment, entitled *This Common Inheritance*. This very thorough and well-presented document gives a detailed analysis of all areas deserving environmental concern, but gives few concrete plans for the future. However, as a general policy document it may be used to move government thinking more towards considering the environmental implications of its actions during the 1990s.

So what of the future?

There are a number of changes that are bound to come during the remainder of this century and the beginning of the next. Other developments are likely.

From trends that have already started either in the UK or abroad, or from policies that are known to be under consideration, one can discern current thinking. All that is needed is a change in public opinion or in consumer behaviour for such changes to be implemented on a larger scale. During the 1980s a swing in public opinion in what was then West Germany, resulting in the election of some Green MPs to the Bundestag, effectively made that country the leader of the European green lobby.

All new cars have to be fitted with catalytic converters from the beginning of 1993. The taxation of cars will become more progressive, so that large cars attract a greater tax burden than small ones, as is already the case to some extent in France and Italy. However, the taxation of cars may ultimately become much more sophisticated, calculated on the basis of a car's overall environmental rating rather than on the size of its engine. This will probably be co-ordinated at EC level, so as to prevent the European market for cars becoming too divided along national lines.

Private motorized traffic will be excluded from many town and city centres. Such schemes exist in several European urban centres and our own pedestrianized zones are already a move in this direction. Such a policy could be introduced very soon in smaller towns and cities such as Norwich, Cambridge, Gloucester or Aberdeen, but it is probably even more important in central London, where traffic could soon grind to a complete halt. Urban design may well change in the future as a result of such measures and we may see a return to more compact, less car-dependent towns.

By the mid-1990s the first alternative vehicles will have appeared. The city centre traffic bans may in some cases be waived for electric vehicles, which would give these a real boost. Several companies have begun already to offer electric versions of their smaller models. Although they are expensive, Fiat's Panda Elettra, Peugeot's electric 205 and the VW Golf CitySTROMer will pick up some sales. Next we will see production of the hybrid cars, as shown by VW and Audi among others. These are specifically designed in the event of a ban on petrol and diesel cars from city centres.

After this, vehicles designed from the outset to be powered by electricity will enter the market. These will probably not be cheap, but they may attract tax incentives in order to make them more competitive with conventional cars. We also have to consider the problems of electricity generation, although charging the cars overnight when power stations overproduce anyway, coupled with comprehensive energy savings in the home and in industry, may largely cover the increased demand. Cleaner power stations and increasing use of alternative energy sources such as wind, sun and water would make this even more realistic. A mixture of energy sources is really needed, as part of the problem is our reliance on too few sources of energy.

Road pricing is also likely to be introduced during this period, matching the cost of using certain roads with fluctuations in demand. This may be coupled with the implementation of certain parts of the Prometheus programme (see page 102). In the US we may see a great increase in the use of methanol as a fuel, while legislation there, especially in California, will lead to a rapid increase in electric vehicle technologies and their sales.

Some governments will begin to put pressure on employers to locate nearer to their employees or to points that can easily be reached by public transport. Commuting by private car will be put under severe pressure and the British company car as a perk will be eroded or disappear altogether. Alternatively, the UK government may adopt the recent suggestion from a report by Dr Madsen Pirie (1990) of the Adam Smith Institute, which suggests limiting company car tax concessions to electric vehicles. Although appealing in many ways (pollution would shift away from urban centres), it would not solve the major problem of traffic congestion.

Alternative private transport systems, such as bicycles and HPVs, will increase in popularity as some roads become quieter and safer. Bicycle facilities should then obviously also be improved, although it is unlikely that the bicycle will play such an important role in future transport systems in the UK as it does in countries like the Netherlands or Denmark.

Around the millennium we may also see the first of the new lightweight high performance engines using two-stroke technology or similar systems, which will make lightweight cars more viable. True lightweight cars, if they come, will probably have to wait until the early 2000s, while hydrogen-powered cars for the general public will be further away, even if their problems are ever solved. Despite the fact that several producers, such as Toyota, Mercedes-Benz, Cadillac and Mazda, are developing 12-cylinder engines, such engines are likely to disappear before the end of the decade as they are anything but fuel-efficient (this may be reinforced by CAFE-style legislation in Europe and Japan, or by punitive taxation of the largest engines) and wasteful of materials. By the turn of the century most diesel- and petrol-engined cars will still have 4-cylinder engines, with some 3-cylinder engines in the increasing number of mini cars. Luxury cars will have no more than a V6.

The car will become less omnipresent, as its role as a commuting vehicle will largely disappear. It may become more of a leisure vehicle, which was, after all, its original role. These cars will be quite different from those of today. They will almost certainly be lighter, have a longer life expectancy, and many of them may be electrically powered. Many people may even have solar panels on their roofs to help recharge their cars and take pressure off the national grid. This is already the case in parts of Switzerland. The leisure features of electric vehicles can easily include the increasingly popular four-wheel drive option by fitting one electric motor to each wheel. This may reduce the popularity of off-road jeep-type vehicles for road use, as they are fairly inefficient and heavy as road cars.

The changing role of the car need not spell gloom for all manufacturers, as its new status as a leisure tool could lead to an explosion of market niches, in which the emphasis would be firmly on the individuality of the product. Nonetheless, some manufacturers may well go out of business, as their facilities and practices no longer match the requirements of the market, while longer life expectancies and restrictions on car use may lead to a depressed market and overcapacity. As Fritjof Capra puts it in *The Turning Point*:

> Many of the large corporations are now obsolete institutions that lock up capital, management and resources but are unable to adapt their functioning to changing needs. A well-known example is the automobile industry, which is unable to adjust to the fact that the global limitations of energy and resources will force us to drastically restructure our transportation system, shifting to mass transit and to smaller, more durable cars.

New small and innovative producers may emerge, especially in the new segments for electric vehicles and lightweight cars. Some job losses are inevitable, although these may be partly absorbed by a declining workforce in many industrialized countries and by the growth of the new environmental industries responsible for cleaning up the mess.

This is a brief scenario, spanning some two to three decades. This is far too long for the environment to wait. Climatic change has already started and, if many experts are right, will change further in the near future. Can we really wait another 10 to 20 years before we take drastic action? Many people both inside and outside the industry believe that some miraculous technical solution to all our environmental problems is just around the corner. We must not be lulled into a false sense of security by this 'technofix' scenario; it will not happen this time. Any technical solution will merely postpone the inevitable. What is needed is a radical change in attitude, which will motivate people to take a global view of the human population as part of the Earth, inextricably linked to its destiny.

Further reading

Some of the topics discussed are covered in greater detail in the works listed below.

Ballantine, R. (1988), *Richard's New Bicycle Book*, Oxford, The Oxford Illustrated Press.

Banks, R. (ed) (1989), *Costing the Earth*, London, Shepheard Walwyn – CIT.

Bendixson, T. (1977), *Instead of Cars*, London, Pelican.

Boyle, S. & J. Ardill, (1989), *The Greenhouse Effect*, Sevenoaks, Hodder & Stoughton.

Burgess Wise, D. (1973), *Steam on the Road*, London, Hamyln.

Capra, F. (1982), *The Turning Point*, London, Fontana.

Department of the Environment (1990), *The Common Inheritance: Britain's Environmental Strategy*, London, HMSO.

Department of Transport, *New Car Fuel Consumption: The official figures*, London, Central Office of Information. Regularly updated.

Elkington, J. & J. Hailes, (1989), *The Green Consumer Guide*, London, Gollancz.

Elkington, J. & T. Burke, (1989), *The Green Capitalists*, London, Gollancz.

Energy 2000, Report to the World Commission on Environment and Development, London, Zed Books.

Evans, R.J. (1985), *Steam Cars*, Aylesbury, Shire Publications.

Ewers, W. (1977), *Solar Energy*, Dorchester, Prism Press.

Georgano, G.N. (ed) (1986), *The Complete Encyclopaedia of Motorcars 1885-1986*, London, Ebury Press.

Hamer, M. (1987), *Wheels within Wheels: a Study of the Road Lobby*, London, Routledge & Kegan Paul.

HMSO (1990), *Transport Statistics of Great Britain*, London.

Howarth, A. (1987), *Africar: the Development of a Car for Africa*, Lancaster, The Ordinary Road.

International Energy Agency (1984), *Fuel Efficiency of Passenger Cars*, Paris.

Joos, E. (1990), '"The Zurich Model", light transit to combat congestion', in *Public Transport International*, 3/1990.

Lowe, M.D. (1989), *The Bicycle: Vehicle for a Small Planet*, Worldwatch Paper 90, Washington DC.

Meetham, A.R. (1981), *Atmospheric Pollution, its History, Origins and Prevention*, Oxford, Pergamon Press.

Muto, S. (1985), *Automobile Aerodynamics: Car Styling Special Edition*, Tokyo, San'ei Shobo.

Netherlands Ministry of Housing, Physical Planning and Environment (1989), *National Environmental Policy Plan*, The Hague, SDU Publishers.

National Society for Clean Air (1990), *Clean Cars: How to Choose One*, Brighton, NSCA.

Norbye, J.P. (1977), *Streamlining and Car Aerodynamics*, Blue Ridge Summit PA, TAB Books.

Parkin, S. (1989), *Green Parties: an International Guide*, London, Heretic Books.

Parr, A. (1991), *Sandyachting*, Llandysul, Gomer Press.
Pirie, M. (1990), *Green Machines*, London, Adam Smith Institute.
Renner, M. (1988), *Rethinking the Role of the Automobile*, Worldwatch Paper 84, Washington DC.
Shacket, S. (1981), *The Complete Book of Electric Vehicles*, Northbrook, Ill., Domus Books.
Society of Motor Manufacturers and Traders:
(1989), *World Automotive Statistics*, London.
(1990), *How to Drive Clean and Save Money*, London.
(1990), *The Motor Vehicle and the Environment*, London.
(1990), *Public Service Vehicles: Buses and Coaches on the move in Britain*, London.
Ware, C. (1982, updated 1987), *Durable Car Ownership*, Bath, The Morris Minor Centre.
Williams, H. (1991), *Autogeddon*, London, Jonathan Cape.

The Open University Energy and Environment Research Unit Publications:
Cousins, S. (1984), *Opportunities for Automobile Fuel Economy arising from new Technology and Design: Report 046*, Milton Keynes.
Hughes, P. & S. Potter (1989), *Routes to Stable Prosperity: Report 061*, Milton Keynes.
(1990), *Vital Travel Statistics*, Milton Keynes.
Potter, S. (1990), *The Sierra Set Rolls On . . .*, Milton Keynes.
Hughes, P. (1990), *Transport and Pollution: What's the Damage?*, paper presented to LTT conference 27 March 1990.

Open University Courses: T234 Environmental Control and Public Health (1985), 2nd level course; S326 Ecology (1986), 3rd level course; U206 Environment (1990), 3rd level course.

Useful addresses

Air Resources Board, PO Box 2815, 1102 Q Street, Sacramento, CA 95814, USA.

British Federation of Land and Sand Yacht Clubs, Mike Hampton, 23 Piper Drive, Long Whatton, Loughborough, Leicestershire, LE12 5DJ.

British Human Power Club, John Kingsbury, 22 Oakfield Road, Bourne End, Bucks., SL8 5QR.

CleanAir, 10880 Wilshire Blvd., Suite 914, Los Angeles.

Clean Air Transport AB, Odinsgatan 11, S-411 03, Goteborg, Sweden.

Electric Vehicle Association of Great Britain Ltd (EVA), 13 Golden Square, Piccadilly, London W1R 3AG.

Electric Vehicle Council, 1111 19th Street NW, Washington DC 20036, USA.

Energy and Environment Research Unit, Faculty of Technology, The Open University, Milton Keynes, MK7 6AA. Set up in the mid 1970s as an interdisciplinary group to study and campaign on behalf of environmentally benign energy systems for the 21st century and beyond. This organization has produced a number of useful publications on road vehicles and their environmental impact.

Environmental Protection Agency, 401 M Street South West, Washington DC 20460, USA.

Environmental Transport Association (ETA), 15a, George Street, Croydon, CR0 1LA. This organization is aimed at the motorist with an environmental conscience; it offers recovery and roadside assistance like the AA and RAC (but cheaper), but unlike these organizations is not part of the road lobby. It also caters for bicycles, but charges extra for roadside assistance cover for cars older than 10 years (thus promoting premature scrapping of cars?). A good initiative, nonetheless.

European Electric Road Vehicle Association, Place du Trone/Troonplaats 1, B-1000 Brussels, Belgium.

Friends of the Earth (FoE), 26-28 Underwood Street, London N1 7JU. One of the three main general environmental organizations. It deals with all types of threat to the environment in a variety of ways. Particularly strong on green consumerism, including those aspects affecting traffic and vehicles.

General Motors Environmental Activities, 30400 Mound Road, Warren, Michigan 48090-9015, USA.

Greenpeace, 30-31 Islington Green, London N1 8XE. Large international lobbying and campaigning body dealing with all aspects of the environment, including vehicles and transport issues. Noted for its non-violent and widely publicized protests, which have proved effective on several occasions.

International Human-Powered Vehicle Association (IHPVA), PO Box 51255, Indianapolis, IN 51255, USA.

Los Angeles Department of Water and Power, Room 1129, PO Box 111, Los Angeles 90051-0100, USA.

National Society for Clean Air and Environmental Protection, 136 North Street, Brighton, BN1 1RG. This organization captured the public imagination with its active campaigning against lead in the environment. Since the rapid increase in the availability and use of unleaded petrol, it has become concerned with wider issues of environmental pollution, particularly from cars.

Society of Motor Manufacturers and Traders (SMMT), Forbes House, Halkin Street, London SW1X 7DS. This organization represents the car producers and importers in the UK. It does take an interest in environmental matters and has produced a number of publications setting out the producers' position on various green topics.

Transport 2000, Walkden House, 10 Melton Street, London NW1 2EJ. Its origins are as a railway promotion group, but it has become the principal lobbying organization concerned about the effects of transport on the environment. It produces and promotes a number of useful publications and runs regular conferences highlighting problems of transport and environment.

Worldwide Fund for Nature (WWF), Panda House, Weyside Park, Godalming, Surrey GU7 1XR. This general environmental organization used to be known as the World Wildlife Fund. Its new name reflects more closely its general conservation aims. Its interest in issues relating to traffic and environment is illustrated by its close links with the Environmental Transport Association.